The Chua Lectures
From Memristors and Cellular Nonlinear Networks to the Edge of Chaos

Vol. I. Memristors
New Circuit Element with Memory

WORLD SCIENTIFIC SERIES ON NONLINEAR SCIENCE

Editor: Leon O. Chua
University of California, Berkeley

Series A. MONOGRAPHS AND TREATISES*

Volume 86: Elements of Mathematical Theory of Evolutionary Equations in Banach Spaces
A. M. Samoilenko & Y. V. Teplinsky

Volume 87: Integral Dynamical Models: Singularities, Signals and Control
D. Sidorov

Volume 88: Wave Momentum and Quasi-Particles in Physical Acoustics
G. A. Maugin & M. Rousseau

Volume 89: Modeling Love Dynamics
S. Rinaldi, F. D. Rossa, F. Dercole, A. Gragnani & P. Landi

Volume 90: Deterministic Chaos in One-Dimensional Continuous Systems
J. Awrejcewicz, V. A. Krysko, I. V. Papkova & A. V. Krysko

Volume 91: Control of Imperfect Nonlinear Electromechanical Large Scale Systems: From Dynamics to Hardware Implementation
L. Fortuna, A. Buscarino, M. Frasca & C. Famoso

Volume 92: Modeling, Analysis and Control of Dynamical Systems with Friction and Impacts
P. Olejnik, J. Awrejcewicz & M. Fečkan

Volume 93: Coupled Phase Locked Loops: Stability, Synchronization, Chaos and Communication with Chaos
V. V. Matrosov & V. D. Shalfeev

Volume 94: Chaos in Nature (2nd Edition)
C. Letellier

Volume 95: Continuous and Discontinuous Piecewise-Smooth One-Dimensional Maps: Invariant Sets and Bifurcation Structures
V. Avrutin, L. Gardini, I. Sushko & F. Tramontana

Volume 96: Frequency-Domain Approach to Hopf Bifurcation Analysis: Continuous Time-Delayed Systems
F. S. Gentile, J. L. Moiola & Guanrong Chen

Volume 97: Toward General Theory of Differential-Operator and Kinetic Models
N. Sidorov, D. Sidorov & A. Sinitsyn

Volume 98: The Chua Lectures: From Memristors and Cellular Nonlinear Networks to the Edge of Chaos
(In 4 Volumes)
Volume I. Memristors: New Circuit Element with Memory
Volume II. Memristors and CNN: The Right Stuff for AI and Brain-Like Computers
Volume III. Chaos: Chua's Circuit and Complex Nonlinear Phenomena
Volume IV. Local Activity Principle: Chua's Riddle, Turing Machine, and Universal Computing Rule 137
L. O. Chua

*To view the complete list of the published volumes in the series, please visit:
http://www.worldscientific.com/series/wssnsa

WORLD SCIENTIFIC SERIES ON NONLINEAR SCIENCE Series A Vol. 98

Series Editor: Leon O. Chua

The Chua Lectures

From Memristors and Cellular Nonlinear Networks to the Edge of Chaos

Vol. I. Memristors

New Circuit Element with Memory

Leon O. Chua

University of California at Berkeley, USA

World Scientific

NEW JERSEY · LONDON · SINGAPORE · BEIJING · SHANGHAI · HONG KONG · TAIPEI · CHENNAI · TOKYO

Published by

World Scientific Publishing Co. Pte. Ltd.

5 Toh Tuck Link, Singapore 596224

USA office: 27 Warren Street, Suite 401-402, Hackensack, NJ 07601

UK office: 57 Shelton Street, Covent Garden, London WC2H 9HE

Library of Congress Cataloging-in-Publication Data

Names: Chua, Leon O., 1936– author.
Title: The Chua lectures : from memristors and cellular nonlinear networks
to the edge of chaos / Leon O. Chua.
Other titles: Lectures. Selections.
Description: New Jersey : World Scientific, [2020] | Series: World scientific series on
nonlinear science. Series A ; Vol. 98 | Contents:
Volume I. Memristors : new circuit element with memory --
Volume II. Memristors and CNN : the right stuff for AI and brain-like computers --
Volume III. Chaos : Chua's circuit and complex nonlinear phenomena --
Volume IV. Local activity principle : Chua's riddle, Turing machine, and
universal computing rule 137.
Identifiers: LCCN 2020030531 | ISBN 9789811215384 (v. 1 ; hardcover) |
ISBN 9789811215407 (v. 2 ; hardcover) | ISBN 9789811215421 (v. 3 ; hardcover) |
ISBN 9789811215445 (v. 4 ; hardcover) | ISBN 9789811215377 (hardcover) |
ISBN 9789811215391 (v. 1 ; ebook) | ISBN 9789811215414 (v. 2 ; ebook) |
ISBN 9789811215438 (v. 3 ; ebook) | ISBN 9789811215452 (v. 4 ; ebook)
Subjects: LCSH: Memristors. | Chaotic behavior in systems. | Computational complexity.
Classification: LCC TK7872.R4 C49 2020 | DDC 621.3815/4--dc23
LC record available at https://lccn.loc.gov/2020030531

British Library Cataloguing-in-Publication Data

A catalogue record for this book is available from the British Library.

For any available supplementary material, please visit
https://www.worldscientific.com/worldscibooks/10.1142/11693#t=suppl

To

Amy

for whom the world is her oyster

About the Author

Leon O. Chua is known worldwide as the father of *nonlinear circuit theory and inventor* of the *memristor*, the *Cellular Neural Network*, and the *Chua Circuit*, the first real-world system proved to exhibit *chaos*. His publications, which span many disciplines in engineering, biology, brain science, physics, and mathematics, have attracted more than 80,000 citations, with an ***h-index* of 125**, as of 8 April 2020.

His body of work has been recognized by numerous prestigious prizes, including the *IEEE Gustav Robert Kirchhoff Award*, the *IEEE Neural Networks Pioneers Award*, and the *Guggenheim Fellow Award*.

In 1971, he had postulated a new *circuit element*, dubbed the ***memristor***, which was fabricated as a working nano-device only in 2008 by Dr. R. Stanley Williams, and his group of ingenious scientists from **HP labs**. It has since been identified as the fundamental building block for designing intelligent brain-like machines.

In recognition of his discovery of the *memristor*, he has been conferred the most prestigious honors by two *unrelated* professional societies that celebrate the *memristor's* fundamental and multi-disciplinary applications, namely, the **2019 IEEE Electronic Device Society Celebrated member Award**, and the **2020 Julius Springer Prize for Applied Physics**, thereby entrenching him in the company of luminaries who have been previous EDS awardees, like the Nobel laureates George Pearson Smith in 2010, Herbert Kroemer in 2011, and Leo Esaki in 2012, respectively, as well as industry giants like Robert Dennard (of Dennard's scaling law) in 2015, and Gordon Moore (of Moore's Law fame) in 2017.

He was awarded 17 *Honorary Doctorates from major European universities and Japan*, and conferred an *Honorary Fellow* of the Institute of Advanced Study at the Technical University of Munich.

His inventions are widely applied in industry and disseminated synergetically at the *Chua Memristor Center* (*CMC*) in Dresden, Germany, and at the *Chua Memristor Institute* (*CMI*) in Wuhan, China. Professor Chua was elected a member of the ***Academia Europaea***, and the Hungarian Academy of Sciences.

Interestingly, he was also elected *Confrerie des Chevaliers du Tastevin* in 2000. When not immersed in Science, he relaxes by searching for Wagner's leitmotifs, musing over Kandinsky's chaos, and contemplating Wittgenstein's inner thoughts.

Contents

List of Videos and Audios is Available at
https://worldscientific.com/doi/10.1142/11693#t=suppl

Inspirational Insights from Professor Leon Chua's Lecture

Kirk Bresniker, HP Labs Chief Architect, HP Fellow

I'm going to have to watch the replay of Professor Leon Chua's lecture #2 on "Everything You Wish to Know about Memristors but are Afraid to Ask", and correct and expand my notes. There is also a blog that summarizes his last lecture. Professor Chua doesn't dawdle when it comes to going through the material in his lecture series.

There were some great takeaways from his lecture that were expressed in just a few words that I'm going to be thinking about for weeks. For instance, with the biomimetic behavior of complex biologic systems modeled with a handful of passive components once the memristor is involved, what kinds of design tools are we going to need to design and manufacture passive circuits with chaotic behavior that need to be designed, manufactured, and then trained? This is one of Professor Chua's timeless axiomatic approach to analysis.

But what really struck me was the simple lesson borrowed from Albert Einstein: "a model must be simple, but not simpler". You'll know your model is too simple when it can't explain the evidence. For Johannes Kepler, it was the error in the Mars orbit, for Einstein it was the Thought Experiment (gedankenexperiment) about the speed of light, and for Professor Chua, it was the tunnel diode, a type of semiconductor that is capable of very fast operation. Having spent three years to master the vacuum tube, a device that controls electric current between electrodes in an evacuated container, Professor Chua was faced with not just the tunnel diode, but more than a dozen novel components that couldn't be explained.

Rather than serial solve them, he took a step back and solved them all systematically with an expanded model. But he then explored the ramifications of that model and that uncovered the missing element, which led to the memristor prediction. It wasn't just that the model was more accurate in its results, it was that it gave a dramatic new insight into the behavior of entire classes of components, even those which had yet to be realized. The comparison and interesting counterpoint of the model proposed by Brian Josephson for his Josephson junction, where at a junction of two superconductors, a current will flow even if there is no drop in voltage, was accurate but did not generate the same kind of systemic insight.

Being able to listen to someone who has not only glimpsed the beautiful structure of the natural world through mathematics but has brought those insights to so many others is inspiring. If you've got a chance to see the remaining lectures live, I highly recommend it!

Preface

This 4-volume *compendium* contains the verbatim hard copies of all color slides from the **Chua Lectures: *From Memristors and Cellular Nonlinear Networks to the Edge of Chaos*.** The Chua Lecture Series was presented at HP in Palo Alto, during the period from September 8 to November 24, 2015. It consists of 12 lectures and can be accessed via YouTube by typing the keywords "*Chua HP Lectures*" in the YouTube search bar. Each lecture consists of 90 minutes, divided into a formal lecture, a discussion session, and an encore of exquisite vignettes that the audience found mesmerizing.

These lectures share some unique features of the classic *Feynman Lectures on Physics*, as much of the materials are presented in the unique style of the author, and many fascinating new concepts are original as they were discovered or invented by the author himself. Unlike most technical books that suffer a notoriously short life span as their novelties are superseded by superior technologies, this series of Chua lectures are intended to never be obsolete, as many concepts and principles introduced are in fact new laws of nature, written in the language of sophomore-level mathematics, thereby providing the sparks and the *elan vital* for initiating and nurturing future concepts and inventions.

To enhance readability, we have enlarged the font size of the texts and symbols of numerous slides from the original lectures. In addition, to extend the timeliness of current research on Memristor theory and devices, we have inserted, in Lecture 4, many exciting new colorful slides from the invited paper "**5 Non-volatile Memristor Enigmas Solved**" in *Applied Physics A* **(2018, 124:563)**.

Following is a synopsis of the 4 volumes comprising the 12 Chua HP Lectures:

Volume I—covers everything that a researcher may want to know about *memristors* but is too timid to ask.

Volume II—shows that *memristors* can be either volatile or non-volatile and effectively proving that synapses are *non-volatile memristors*, while action potentials are generated by locally-active but *volatile memristors*.

Volume III—presents an overview of the fascinating phenomenon called *chaos*, while immersing the audience with the *sights and sound of chaos* from the *Chua Circuit*, invented in 1984 by Leon Chua, and has since become the standard textbook example of chaos exhibited by a 5-element electronic circuit, and not by computer simulations.

Volume IV—surprises the audience with a new law of nature—dubbed the *local activity principle*, as discovered and proved mathematically in 1966 by Leon Chua. In particular, a Corollary of *Chua's local activity theorem*, dubbed the *edge of chaos*, is shown via insightful examples to be the genesis of most complex phenomena, including intelligence, creativity and deep learning. Indeed the maxim proclaiming "*Complexity is impossible without edge of chaos*" is now widely regarded as a *new law of thermodynamics*.

The *edge of chaos* is Mother Nature's tool for overcoming the tyranny of the second law of thermodynamics by providing an escape hatch for *entropy to decrease* over time. Indeed, the *local activity principle*, which is profusely illustrated in the final volume, is widely recognized as a new law of nature, and is identified as the *sine qua non* of all complex phenomena, including life itself.

Acknowledgments

Much of the research featured in this lecture series is supported by multiple research grants from the USA Office of Naval Research, and the Air Force Office of Scientific Research (most recently under grant number FA9550-18-1-0016).

The author would like to thank Dr. Stanley Williams, director of the HP Labs in 2015, for inviting him to present the Chua Lecture Series, and Mr. Richard Lewington from HP Technical Communications for setting up the complex logistics for streaming the lectures to all HP's Laboratories worldwide.

It is a great pleasure for the author to thank Dr. K. K. Phua, Chairman of World Scientific Publishing Company, for his encouragement to publish the entire collection of the one-of-a-kind colorful slides featured in this 4-volume tome. He would also want to thank Ms. Lakshmi Narayanan for her superb editorial leadership in shepherding this unique publishing event.

Last but not the least, the author would like to thank Prof. Hyongsuk Kim from Jeonbuk National University South Korea, as well as Dr. Valery Sbitnev, Dr. Zubaer Mannan and Mr. Muhammad Umraiz, for their invaluable assistance in the generation of most of the colorful slides in this lecture series.

May, 2020

Leon Chua
University of California, Berkeley

Lecture 1

Once over Lightly

The Chua Lectures:
From Memristors
and
Cellular Nonlinear Networks
to the
Edge of Chaos

1-1

Lecture 1
Once Over Lightly

1-2

Part 1

Memristors

1-3

1960 CRISIS IN Electronic Circuit Theory

1-4

Esaki Diode

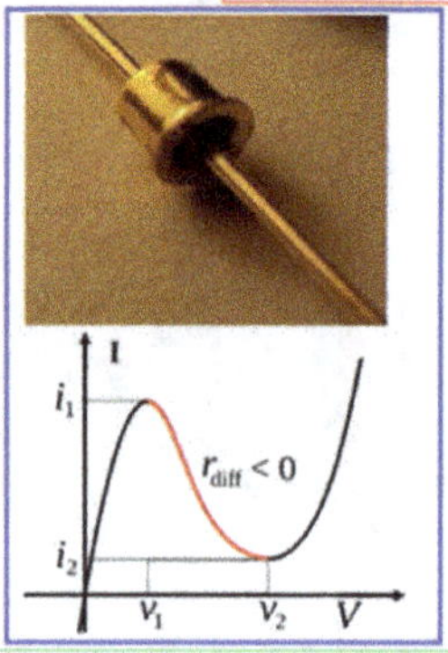

Leo Esaki
Nobel Prize in Physics, 1973

L. Esaki
New Phenomenon in Narrow Germanium p-n junctions
Physics review 109(2):603, 1958

1-5

Simplest *Tunnel Diode* Circuit

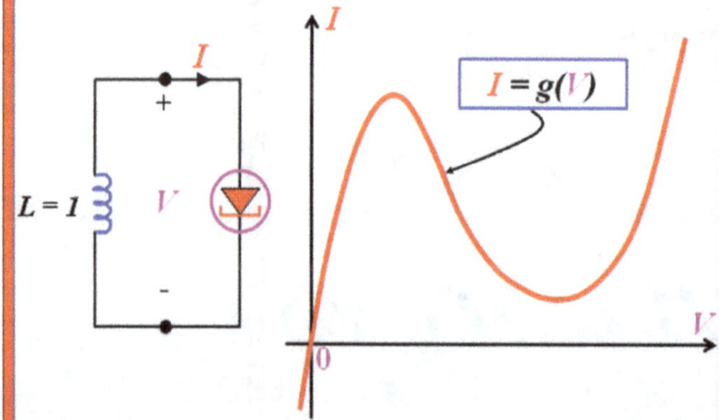

1-6

Lecture 2 1-7

Everything You Wish to Know About Memristors but are Afraid to Ask

Varactor Diode 1-8

Developed by Ramo Wooldridge Corporation patent issued 1961

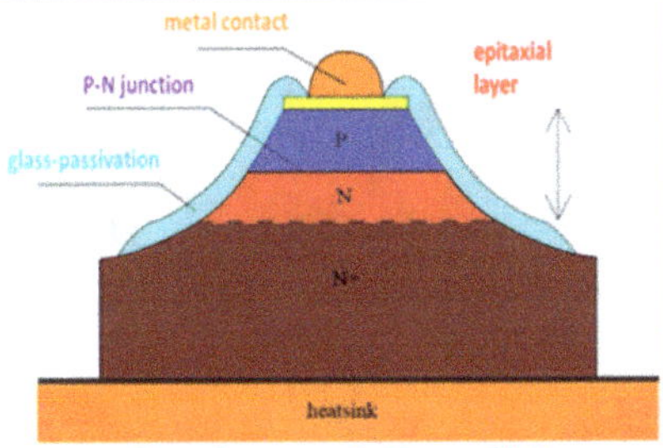

Internal structure of a Varactor Diode

Type	*passive*
Pin configuration	*anode and cathode*

Electronic symbol

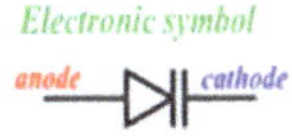

3 Basic Circuit Elements 1-9

Year		Equation
1745		$i(t) = C \frac{dv(t)}{dt}$
1827	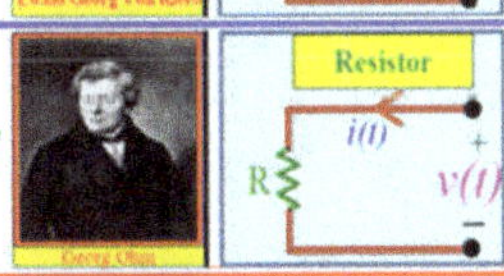	$v(t) = R\, i(t)$
1831		$v(t) = L \frac{di(t)}{dt}$

A Fourth Basic Element

Called the

Memristor

was postulated in **1971**

Leon O. Chua
Memristor : The missing circuit element
IEEE Transactions on Circuit Theory, vol.18, no.5, p.507-519, 1971.

and found in **2008**

D. B. Strukov, G. S. Snider, D. R. Stewart, and R. S. Williams
The Missing Memristor Found
Nature, vol.453, p.80-83, 2008.

1-10

Experimental Definition of the Memristor

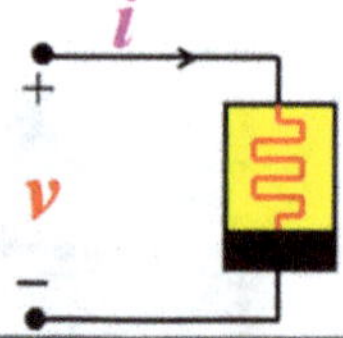

If it's Pinched
It's a
Memristor

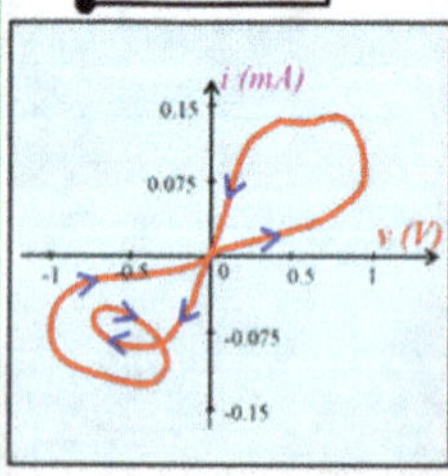

1-11

HP TiO_2 Memristor, Circa 2008

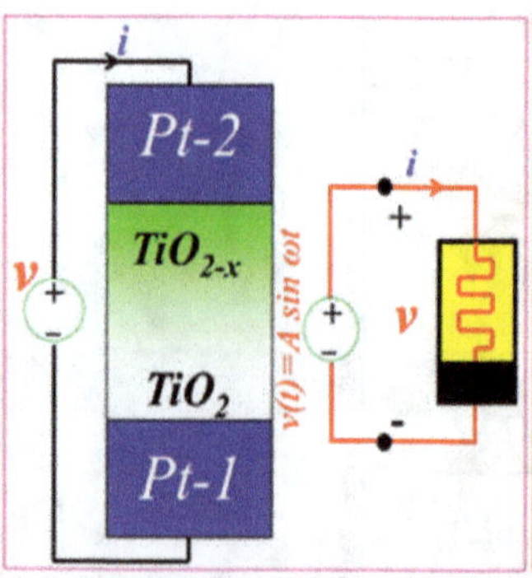

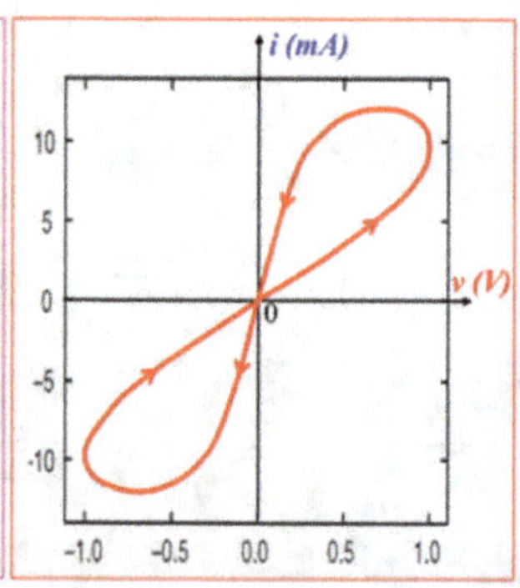

1-12

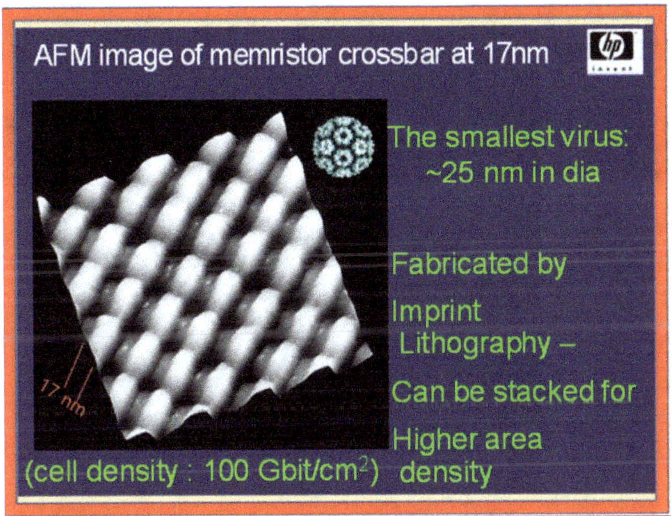
AFM image of memristor crossbar at 17nm
The smallest virus: ~25 nm in dia
Fabricated by
Imprint Lithography –
Can be stacked for
Higher area density
17 nm
(cell density : 100 Gbit/cm^2)

1-13

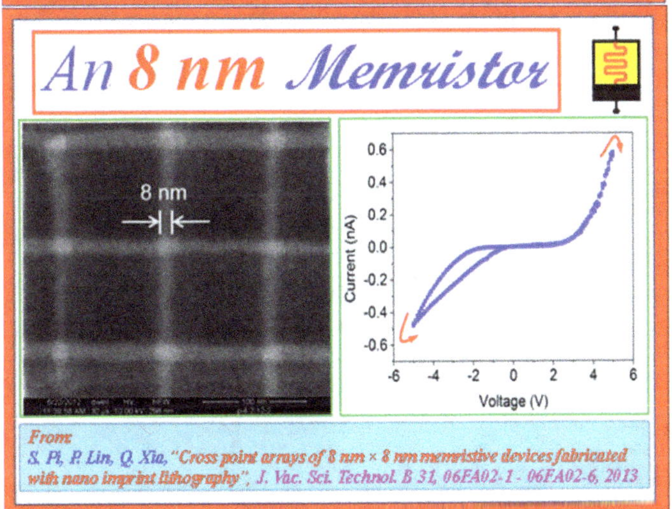
An 8 nm Memristor
8 nm
Current (nA)
Voltage (V)
From:
S. Pi, P. Lin, Q. Xia, "Cross point arrays of 8 nm × 8 nm memristive devices fabricated with nano imprint lithography", J. Vac. Sci. Technol. B 31, 06FA02-1 - 06FA02-6, 2013

1-14

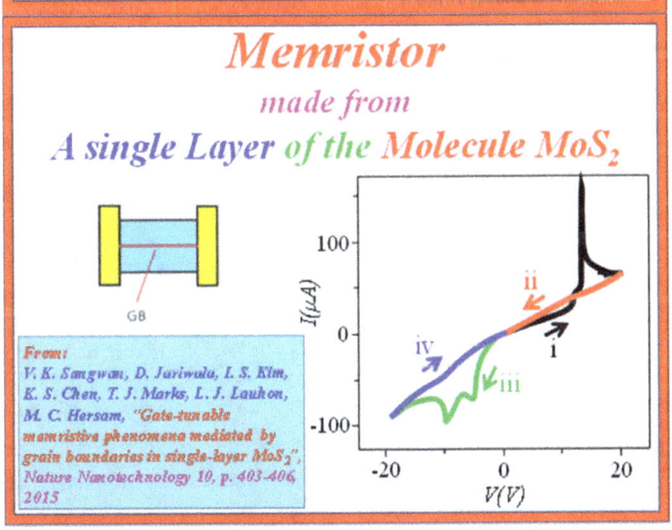
Memristor
made from
A single Layer of the Molecule MoS_2
GB
$I(\mu A)$
$V(V)$
i
ii
iii
iv
From:
V. K. Sangwan, D. Jariwala, I. S. Kim, K. S. Chen, T. J. Marks, L. J. Lauhon, M. C. Hersam, "Gate-tunable memristive phenomena mediated by grain boundaries in single-layer MoS_2", Nature Nanotechnology 10, p. 403-406, 2015

1-15

One-Molecule Memristor

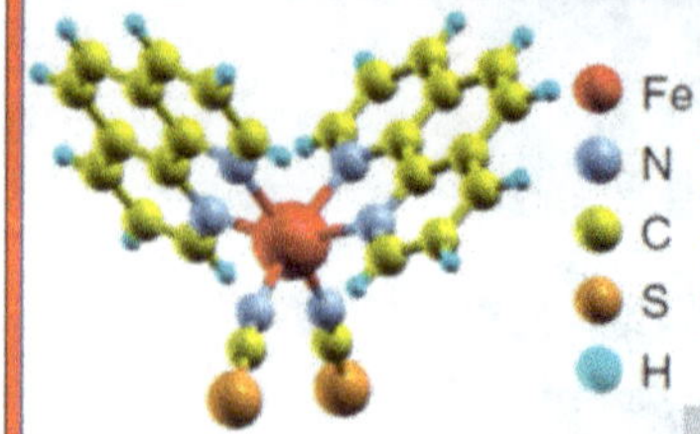

Toshio Miyamachi et al.
" Robust spin crossover and memristance across a single molecule "
Nature Communications 3, 938, p.1-6, 2012

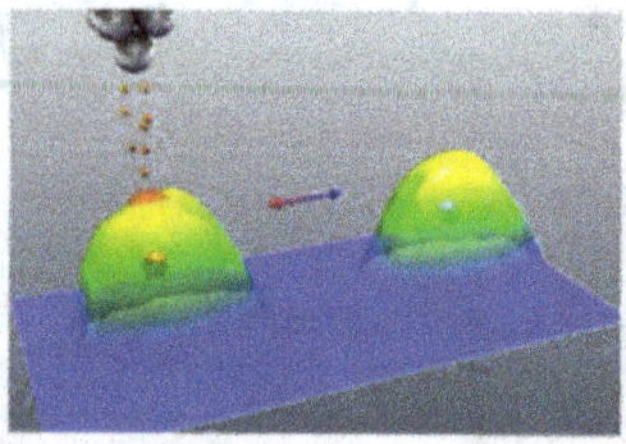

Nature: Molecule Changes Magnetism and Conductance

KIT Press Release No.116, July 2012
Karlsruhe Institute of Technology

1-16

All

Non-volatile Memories

based on

Resistance Switchings

are

Memristors

1-17

Following Non-volatile Memory Devices are

Memristors

- **ReRAMS**
- **Phase Change Memories**
- **MRAMS**
- **Ferro-Electric Memories**
- **CBRAMS**
- **Atomic Switch**

1-18

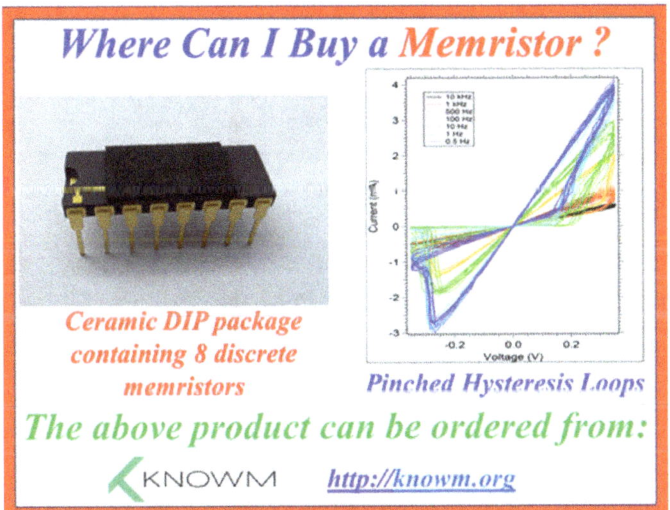
Where Can I Buy a Memristor ?
Current (mA)
Voltage (V)
Ceramic DIP package containing 8 discrete memristors
Pinched Hysteresis Loops
The above product can be ordered from:
KNOWM
http://knowm.org

1-19

1-20

The entire library of congress can be stored in a multi-layer Hp memristor chip
library of Congress
21,218,408 books
10 terabytes (10^12)

1-21

Two *memristors* suffice to compute all Boolean functions

1-22

From:
E. Lehtonen, J.H. Poikonen, and M. Laiho
Electronic Letters, vol. 3, pp. 239, 2010.

Serial CMOS Full Adder

1-23

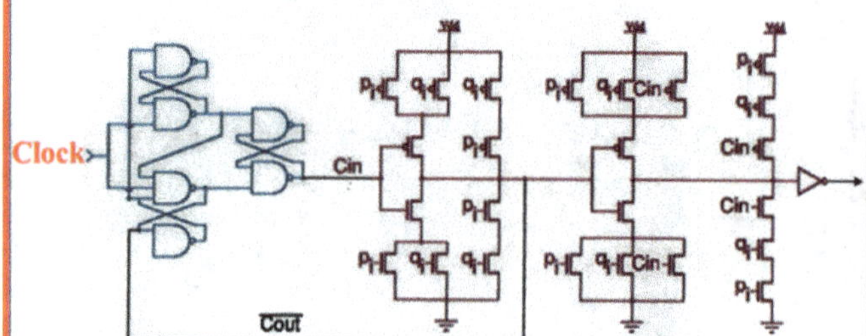

D flip-flop:24 transistors **Full adder: 26 transistors**

Number of Transistors Required : 50

Full Adder Using Memristors

1-24

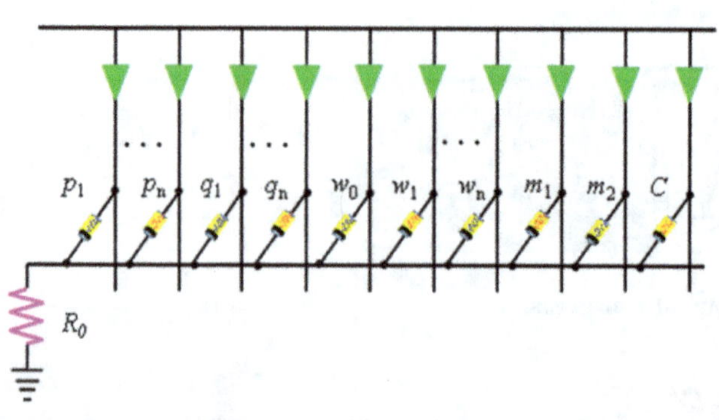

Oscillators

need at least

2

Energy storage Elements

1-25

A One-Memristor Oscillator

1-26

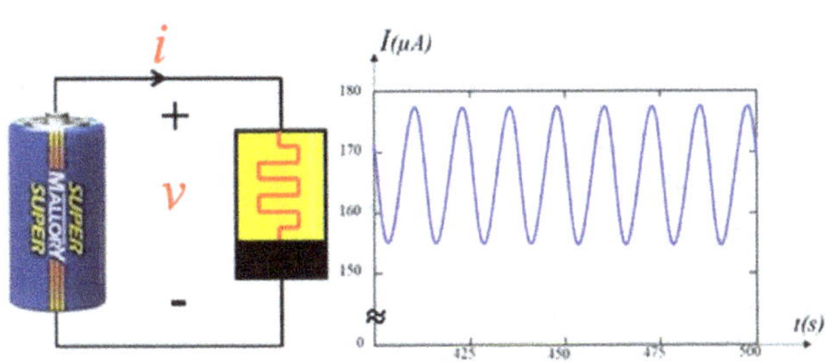

Battery Voltage
E=6.4V

Edge of Chaos Domain
$6.209V < V < 6.38820157073\ V$

Chaos From Chua's Circuit

1-27

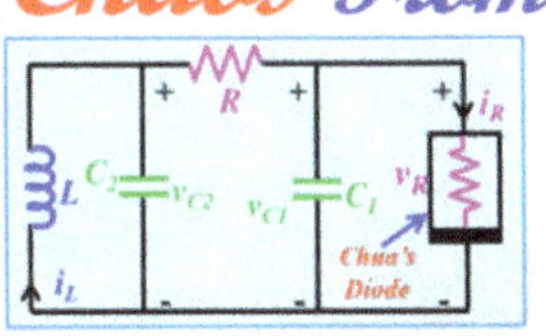

Chua's Circuit

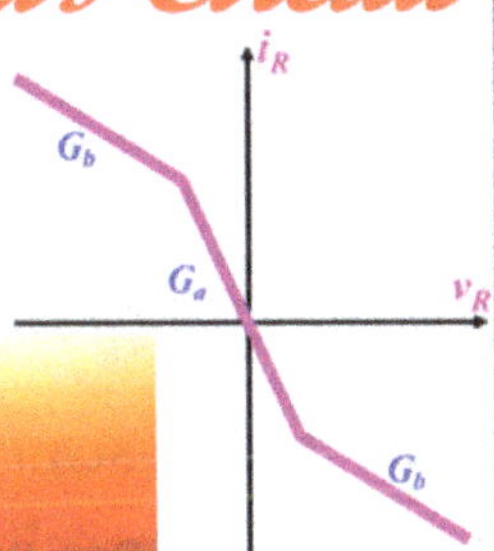

V-I Curve of Chua's Diode

Chua's diode is locally active

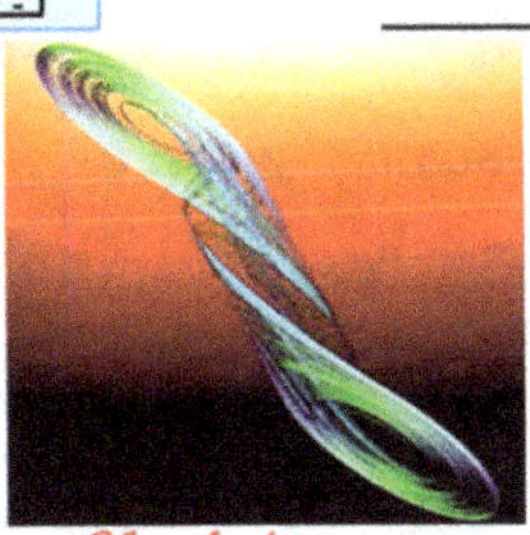

Chua's Attractor

Chaos

need at least

3

Energy Storage Elements 1-28

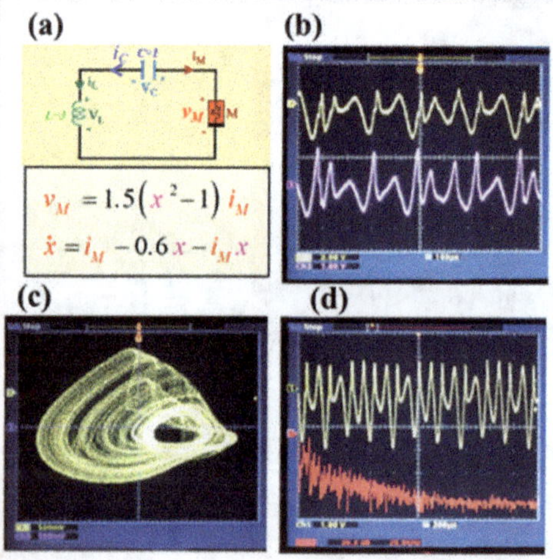

The simplest possible chaotic circuit where the equation describes a Generic Memristor with memristance *$R(x) = 1.5\,(x^2 - 1)$* 1-29

Memristors Can Extend Moore's Law !

Since memristors has both *nonlinearity and dynamics,* and can be scaled down to *sub-nanometer* size, it can increase functionality while reducing the number of devices in hybrid integrated circuits, without shrinking transistors. 1-30

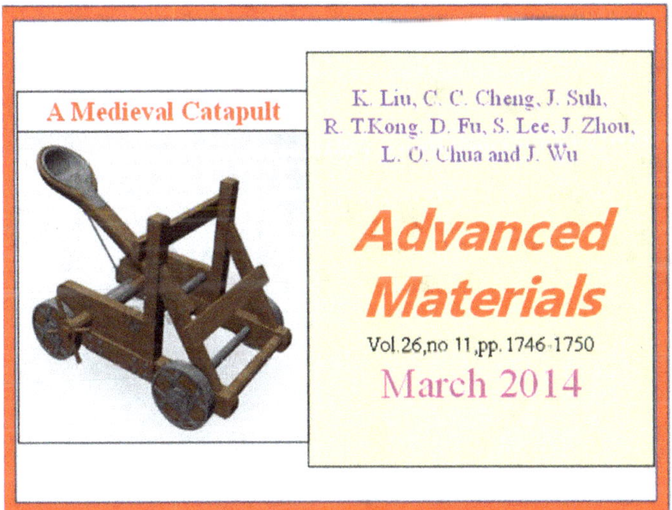
A Medieval Catapult
K. Liu, C. C. Cheng, J. Suh,
R. T.Kong, D. Fu, S. Lee, J. Zhou,
L. O. Chua and J. Wu
Advanced Materials
Vol.26,no 11,pp.1746-1750
March 2014

1-31

Dr. Junqiao Wu with President Obama
at the award ceremony for
Presidential Early Career Awards
for Scientists and Engineers
Whitehouse, April 14, 2014.

1-32

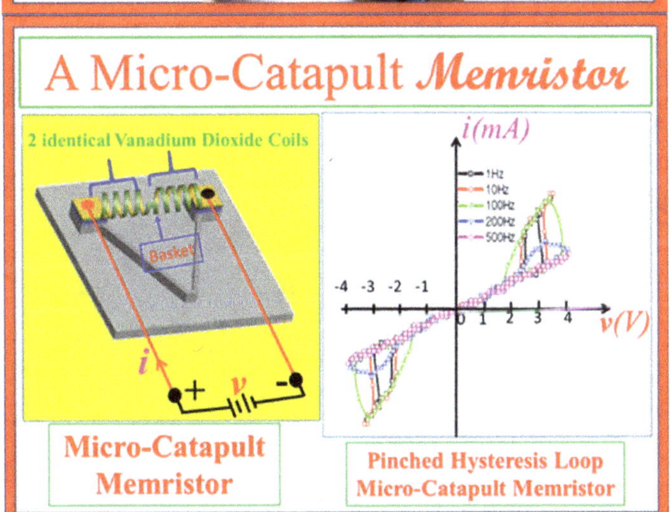
A Micro-Catapult Memristor
2 identical Vanadium Dioxide Coils
Basket
i
+
v
-
Micro-Catapult
Memristor
i(mA)
1Hz
10Hz
100Hz
200Hz
500Hz
-4 -3 -2 -1
0 1 2 3 4
v(V)
Pinched Hysteresis Loop
Micro-Catapult Memristor

1-33

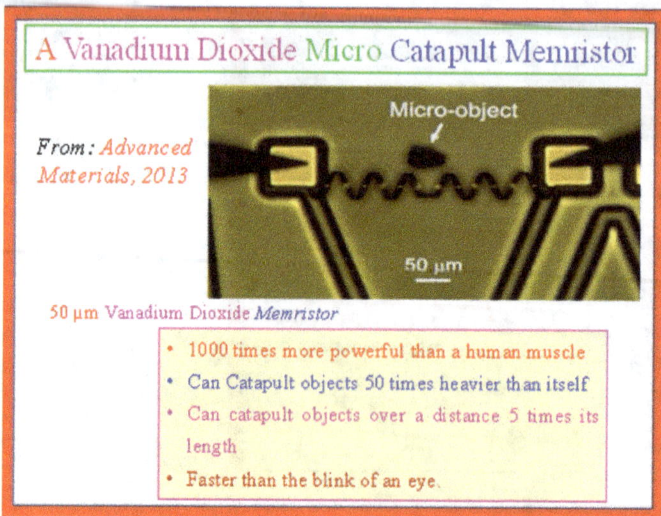
A Vanadium Dioxide Micro Catapult Memristor
From: Advanced Materials, 2013
Micro-object
50 μm
50 μm Vanadium Dioxide Memristor
• 1000 times more powerful than a human muscle
• Can Catapult objects 50 times heavier than itself
• Can catapult objects over a distance 5 times its length
• Faster than the blink of an eye.

1-34

GE90 Turbofan for Boeing 777
Performance of Turbofan for Boeing 777:
Power Density: 9 kW/kg
Rotation Speed: 10,000 rpm
Performance of Micro Catapult Memristor:
Power Density: 39 kW/kg
Rotation Speed: 200,000 rpm

1-35

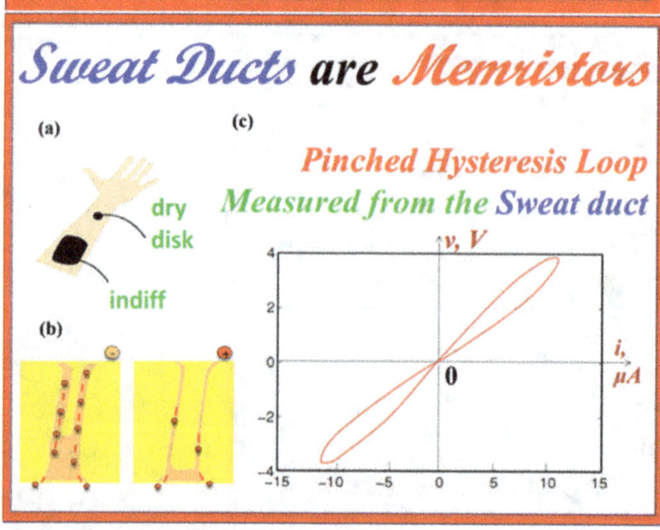
Sweat Ducts are Memristors
(a)
dry
disk
indiff
(b)
(c)
Pinched Hysteresis Loop
Measured from the Sweat duct
v, V
i, μA
0
4
2
0
-2
-4
-15
-10
-5
0
5
10
15

1-36

Our body
is
covered with
Memristors ! 1-37

Memristor
in
Plants 1-38

A. Volkov, C. Tucket, J. Reedus, M. Volkova,
V. S. Markin, and L. Chua

Journal of
Plant Signaling and Behavior
2014

Pinched Hysteresis Loop Measured from the Venus Flytrap 1-39

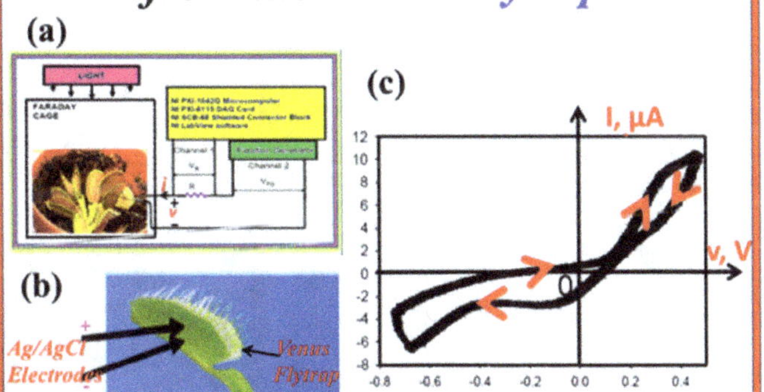

Pinched Hysteresis Loop

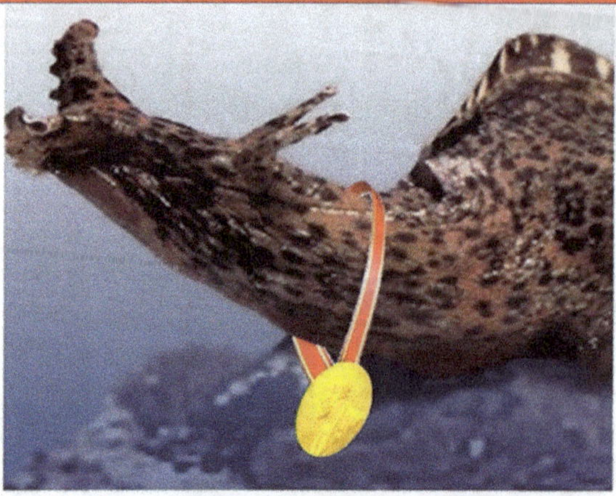

Aplysia with a Nobel Prize Medal

1-40

1-41

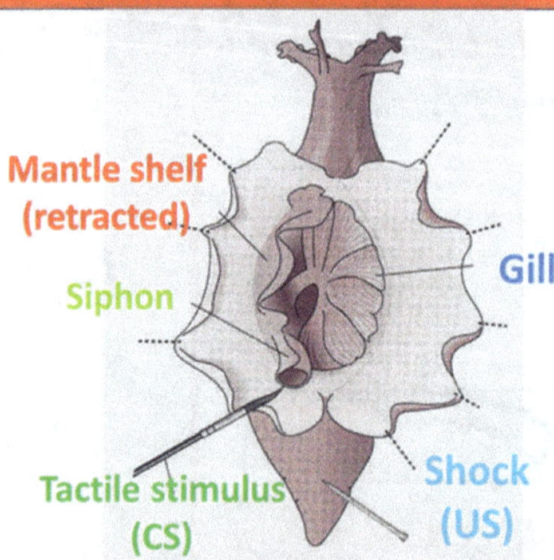

1-42

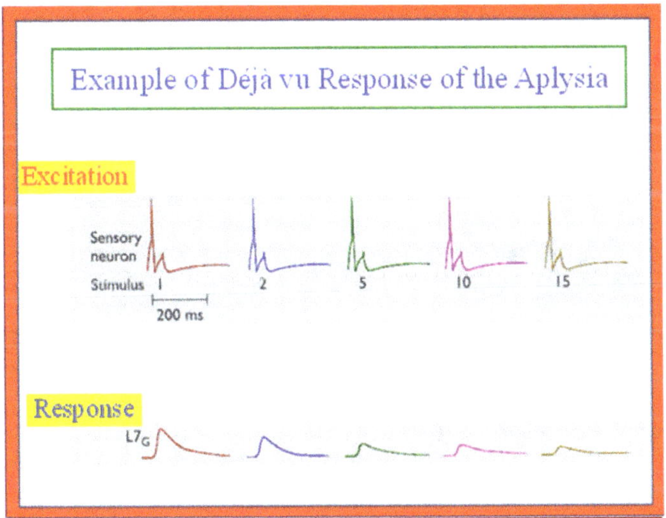
Example of Déjà vu Response of the Aplysia
Excitation
Sensory neuron
Stimulus 1 2 5 10 15
200 ms
Response
L7G

1-43

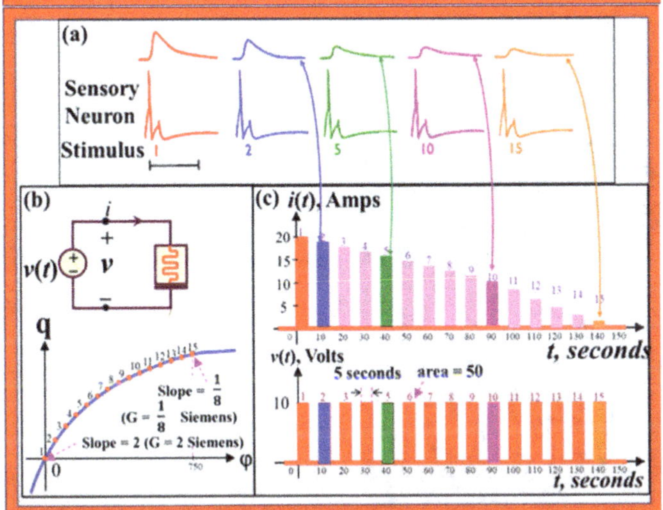
(a)
Sensory Neuron
Stimulus
1 2 5 10 15
(b)
i
v(t)
v
q
Slope = 1/8 (G = 1/8 Siemens)
Slope = 2 (G = 2 Siemens)
φ
(c) i(t), Amps
t, seconds
v(t), Volts
5 seconds
area = 50
t, seconds

1-44

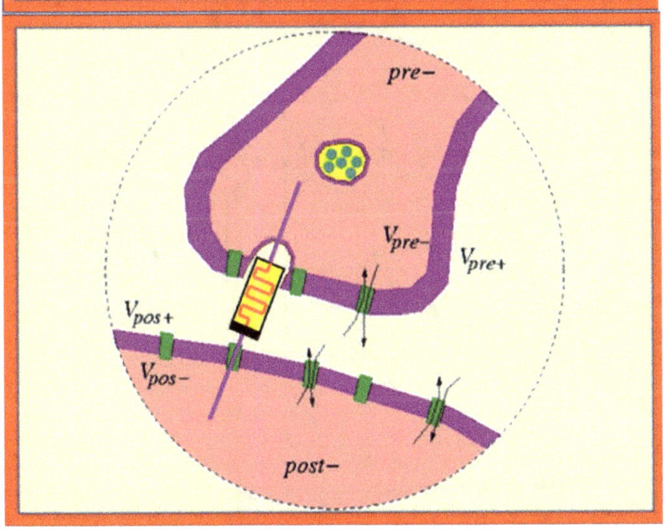
pre−
V_pre−
V_pre+
V_pos+
V_pos−
post−

1-45

Synapses are *Memristors!*

1-46

1-47

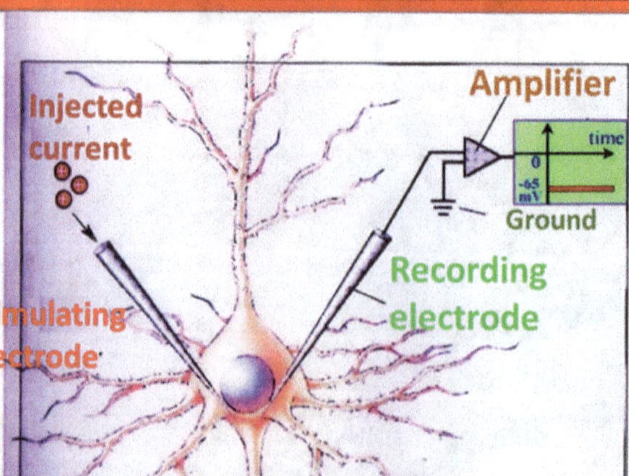

1-48

1961 Nobel Prize in Physiology
Hodgkin- Huxley Nerve Membrane Model
Time-varying Sodium conductance
Time-varying Potassium conductance
Sir A. L. Hodgkin
Sir A. F. Huxley
From
A.L. Hodgkin and A. F. Huxley, "A Quantitative Description of Membrane Current and its Application to Conduction and Excitation in Nerve",
Journal of Physiology, Vol. 117, pp.500-544, 1952

1-49

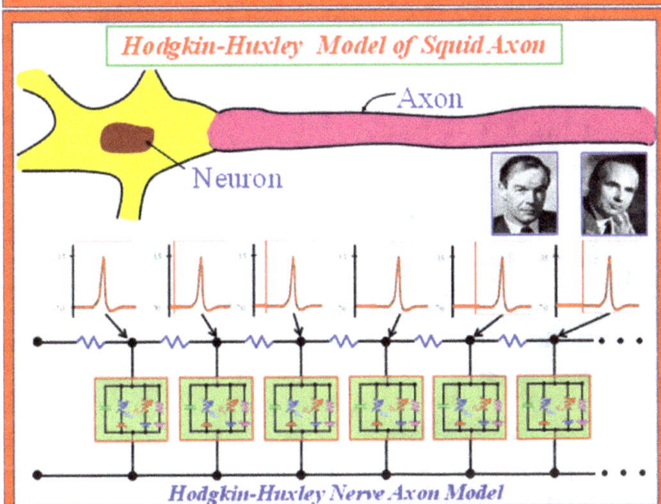
Hodgkin-Huxley Model of Squid Axon
Axon
Neuron
Hodgkin-Huxley Nerve Axon Model

1-50

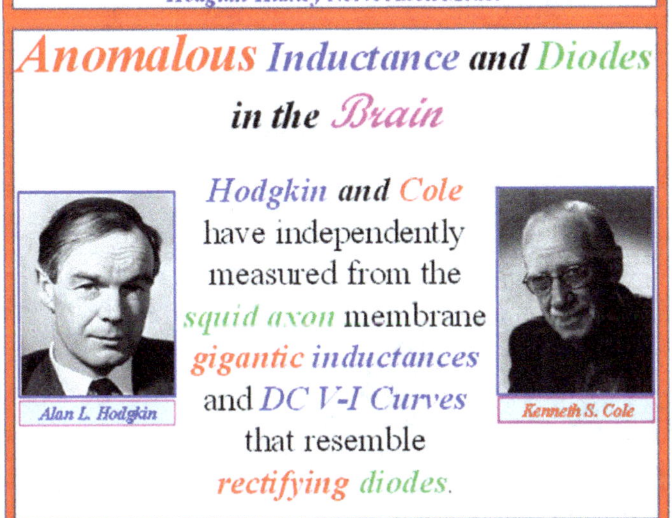
Anomalous Inductance and Diodes in the Brain
Hodgkin and Cole have independently measured from the squid axon membrane gigantic inductances and DC V-I Curves that resemble rectifying diodes.
Alan L. Hodgkin
Kenneth S. Cole

1-51

The suggestion of an inductive reactance anywhere in the system was shocking to the point of being unbelievable.

Kenneth Cole

1-52

1-53

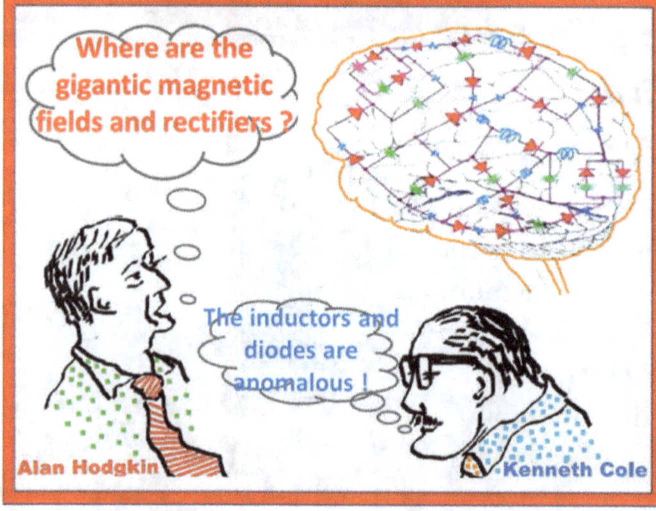

1-54

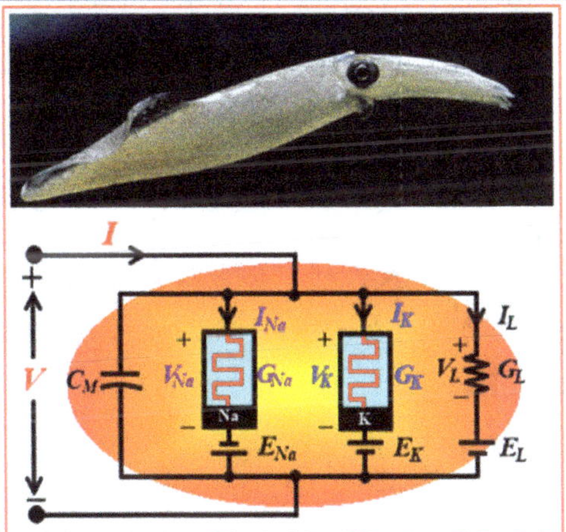

Memristive Hodgkin-Huxley Model

1-55

Axons

are made of

Memristors !

1-56

Brains

are made of

Memristors !

1-57

The Right Stuff

Memristors

are

the right stuff

for making

brainlike

computers

1-58

Part 2

Cellular Nonlinear Networks

1-59

TOSHIBA

Smart Photo Sensor

SPSO2

1-60

CNN Universal Machine

A brain-like image processor chip
which computes elementary
image processing and
recognition tasks

in nano seconds
via
Nonlinear dynamics

1-61

1-62

1-63

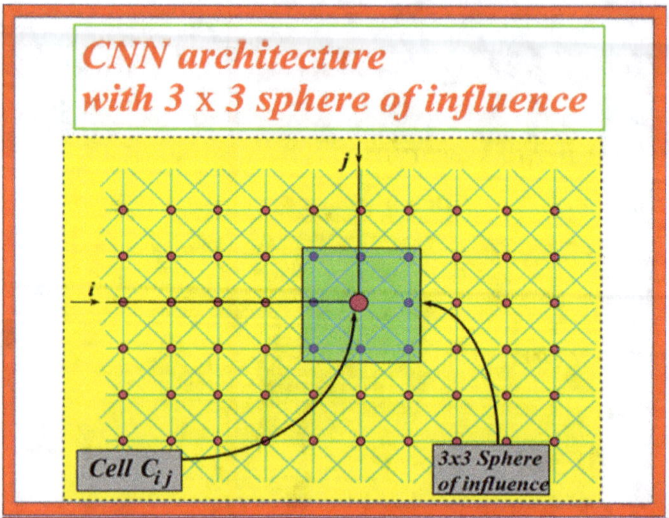
CNN architecture
with 3 x 3 sphere of influence
j
i
Cell C_{ij}
3x3 Sphere
of influence

1-64

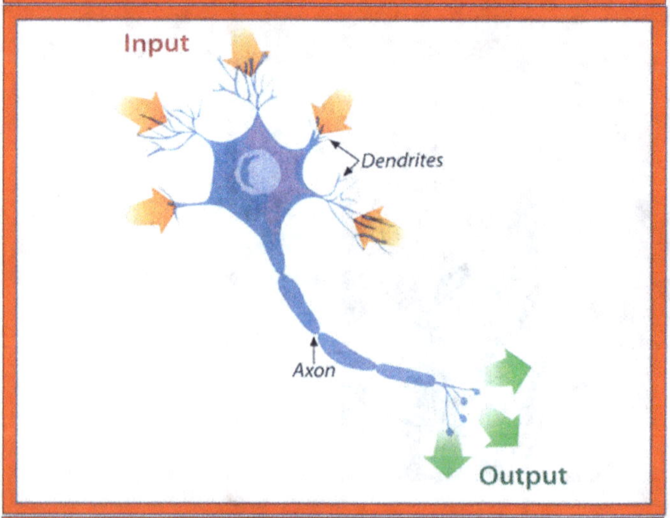
Input
Dendrites
Axon
Output

1-65

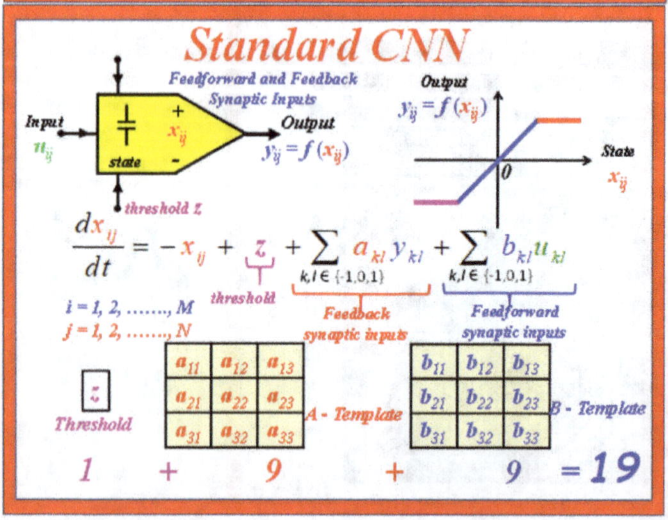
Standard CNN
Feedforward and Feedback
Synaptic Inputs
Input
u_{ij}
state
x_{ij}
Output
$y_{ij} = f(x_{ij})$
threshold z
Output
$y_{ij} = f(x_{ij})$
State
x_{ij}
0
$\frac{dx_{ij}}{dt} = -x_{ij} + z + \sum_{k,l \in \{-1,0,1\}} a_{kl} y_{kl} + \sum_{k,l \in \{-1,0,1\}} b_{kl} u_{kl}$
threshold
Feedback
synaptic inputs
Feedforward
synaptic inputs
i = 1, 2, ..., M
j = 1, 2, ..., N
z
Threshold
a_{11} a_{12} a_{13}
a_{21} a_{22} a_{23}
a_{31} a_{32} a_{33}
A - Template
b_{11} b_{12} b_{13}
b_{21} b_{22} b_{23}
b_{31} b_{32} b_{33}
B - Template
1 + 9 + 9 = 19

1-66

Only 19 Numbers!

The instruction for executing an

Elementary Image-Processing task

is specified by only

19 Real Numbers

Called

"Cloning Templates"

or a

"CNN Gene"

1-67

Templates For Edge Detection CNN

z: -0.5

B:

-1	-1	-1
-1	8	-1
-1	-1	-1

A:

0	0	0
0	2	0
0	0	0

Edge Detection Gene

-0.5	-1	-1	-1	-1	8	-1	-1	-1	-1	0	0	0	0	2	0	0	0	0

1-68

Edge Detection CNN

-0.5	-1	-1	-1	-1	8	-1	-1	-1	-1	0	0	0	0	2	0	0	0	0

1-69

Templates For Corner Detection CNN

z: -8.5

B:

-1	-1	-1
-1	8	-1
-1	-1	-1

A:

0	0	0
0	2	0
0	0	0

Corner Detection Gene

-8.5	-1	-1	-1	-1	8	-1	-1	-1	-1	0	0	0	0	2	0	0	0	0

1-70

Corner Detection CNN

-8.5	-1	-1	-1	-1	8	-1	-1	-1	-1	0	0	0	0	2	0	0	0	0

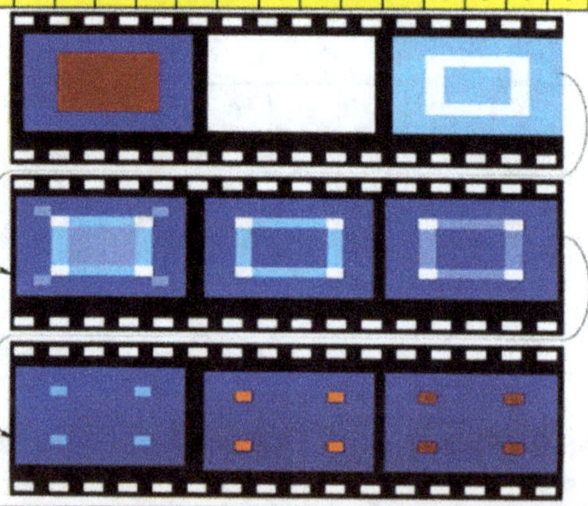

1-71

Corner Detector

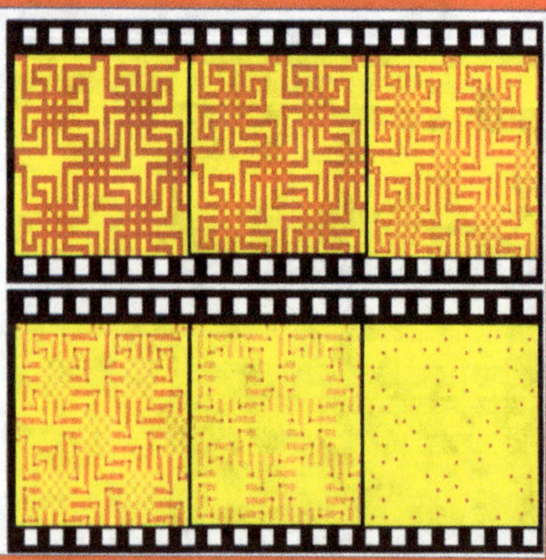

1-72

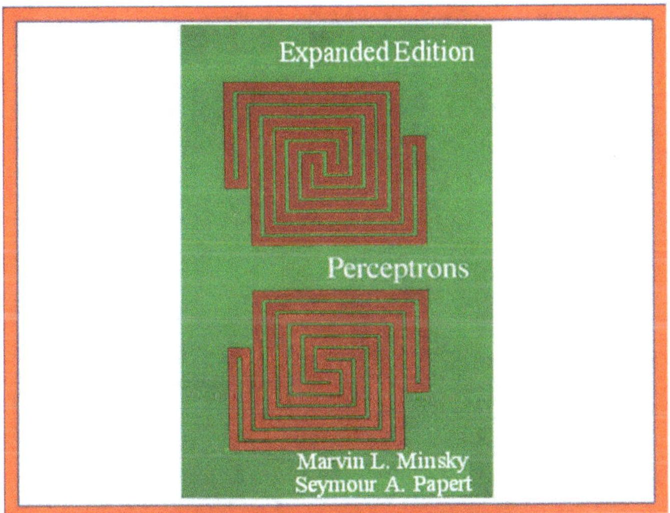
Expanded Edition
Perceptrons
Marvin L. Minsky
Seymour A. Papert

1-73

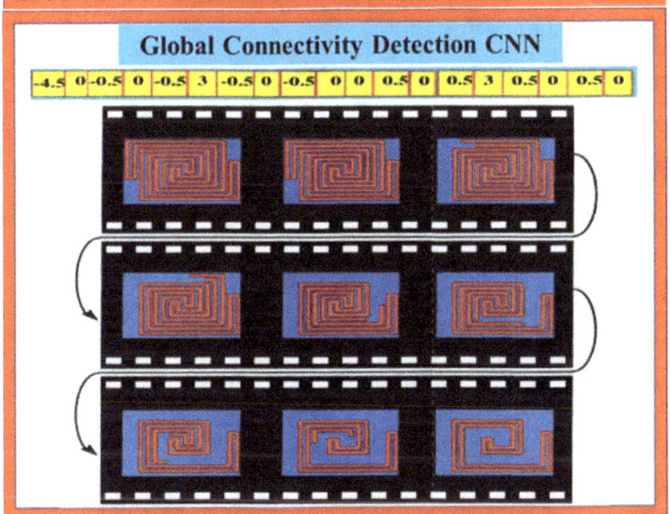
Global Connectivity Detection CNN
-4.5 0 -0.5 0 -0.5 3 -0.5 0 -0.5 0 0 0.5 0 0.5 3 0.5 0 0.5 0

1-74

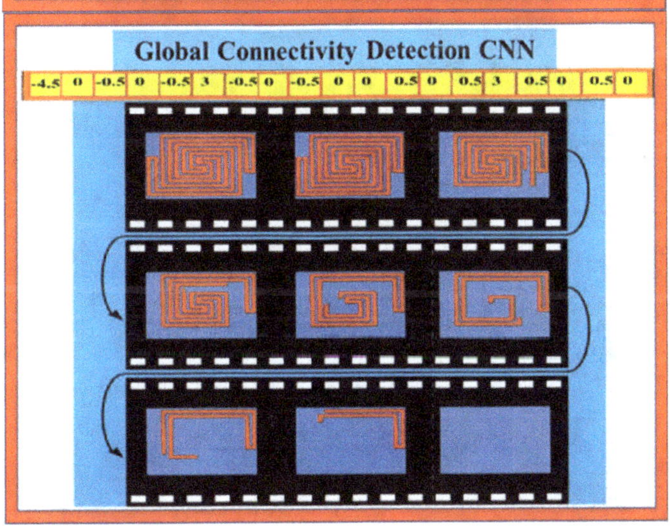
Global Connectivity Detection CNN
-4.5 0 -0.5 0 -0.5 3 -0.5 0 -0.5 0 0 0.5 0 0.5 3 0.5 0 0.5 0

1-75

1-76

Name: **FACE-VASE ILLUSION CNN**

Task Prescription: *Simulate the well-known visual illusion where the following input image is perceived either as two symmetric faces, or as a vase, depending on the initial thought or attention, which is simulated by specifying a small patch of black (red in pseudo-color) pixels inside the object to be picked out. As usual, "red" corresponds to object and "blue" corresponds to background.*

Cloning Template

z: 0

B:

0	0	0
0	-5	0
0	0	0

A:

0	1	0
1	2	1
0	1	0

Boundary Condition

*Fixed: $x_{i^*j^*} = -1$, $u_{i^*j^*} = -1$ (i*j* denotes boundary cells)*

Initial State

$x_{ij}(0)$ = a given binary image

Example 1: Array Size = 200 x 400

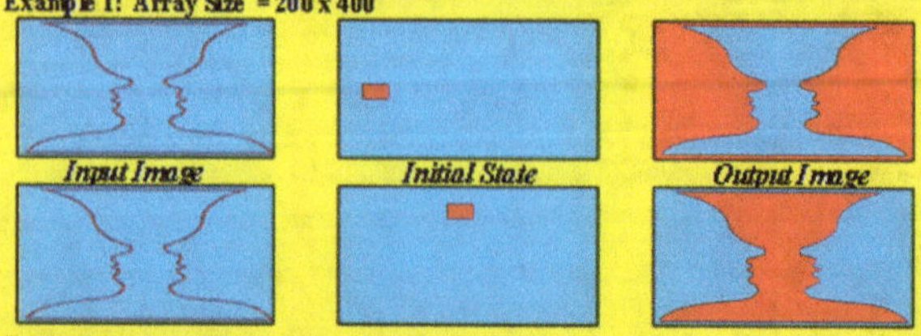

Input Image *Initial State* *Output Image*

Example 2: Array Size = 200 x 400

1-77

Name: **MULLER-LYER ILLUSION CNN**

Task Prescription: *Simulate the well-known illusion where two line segments of identical length but terminated respectively by converging and diverging arrowheads is perceived by humans as if one segment is shorter than the other. For the input image on the left, the CNN must lengthen the upper segment and shorten the lower segment slightly so that that the CNN also exhibits the same illusion.*

Cloning Template

z: -2.8

B:

-0.1	-0.1	-0.1	-0.1	-0.1
-0.1	-0.1	-0.1	-0.1	-0.1
-0.1	-0.1	1.3	-0.1	-0.1
-0.1	-0.1	-0.1	-0.1	-0.1
-0.1	-0.1	-0.1	-0.1	-0.1

A:

0	0	0	0	0
0	0	0	0	0
0	0	1.3	0	0
0	0	0	0	0
0	0	0	0	0

Boundary Condition

*Fixed: $x_{i^*j^*} = 0$, $u_{i^*j^*} = 0$ (i*j* denotes boundary cells)*

Initial State

$x_{ij}(0) = u_{ij}$ = input image

Example 1: Array Size = 44 x 44

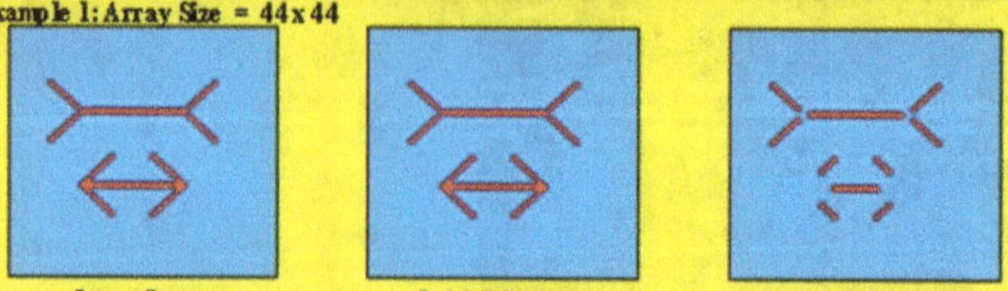

Input Image *Initial State* *Output Image*

1-78

Name: **MULLER-LYER ILLUSION CNN**

Task Prescription: *Simulate the well-known illusion where two line segments of identical length but terminated respectively by converging and diverging arrowheads is perceived by humans as if one segment is shorter than the other. For the input image on the left, the CNN must lengthen the upper segment and shorten the lower segment slightly so that that the CNN also exhibits the same illusion.*

Cloning Template

z: -2.8

B:

-0.1	-0.1	-0.1	-0.1	-0.1
-0.1	-0.1	-0.1	-0.1	-0.1
-0.1	-0.1	1.3	-0.1	-0.1
-0.1	-0.1	-0.1	-0.1	-0.1
-0.1	-0.1	-0.1	-0.1	-0.1

A:

0	0	0	0	0
0	0	0	0	0
0	0	1.3	0	0
0	0	0	0	0
0	0	0	0	0

Boundary Condition

*Fixed: $x_{i^*j^*} = 0$, $u_{i^*j^*} = 0$ (i*j* denotes boundary cells)*

Initial State

$x_{ij}(0) = u_{ij}$ = input image

Example 1: Array Size = 44 x 44

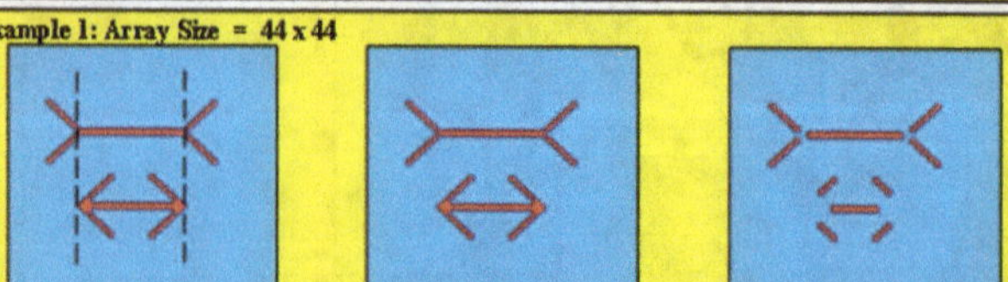

Input Image *Initial State* *Output Image*

Connecting Smart Photo Sensor to A PC

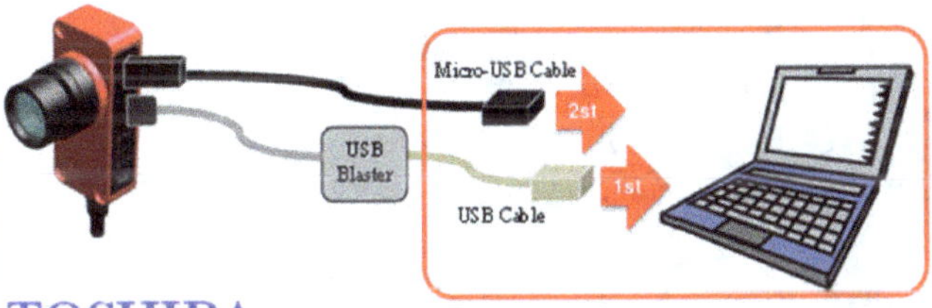

TOSHIBA
SPSO2

1-79

CNN Chip is Very Fast

A typical ***instruction*** *can be implemented on a* ***CNN chip*** *in less than the time for* ***light*** *to travel 1 foot.*

1-80

The ***CNN universal chip*** *is so fast that you can detect a* ***bullet*** *in flight, and have* ***sufficient time*** *to program another* ***bullet*** *to collide with it.*

1-81

1-82

1-83

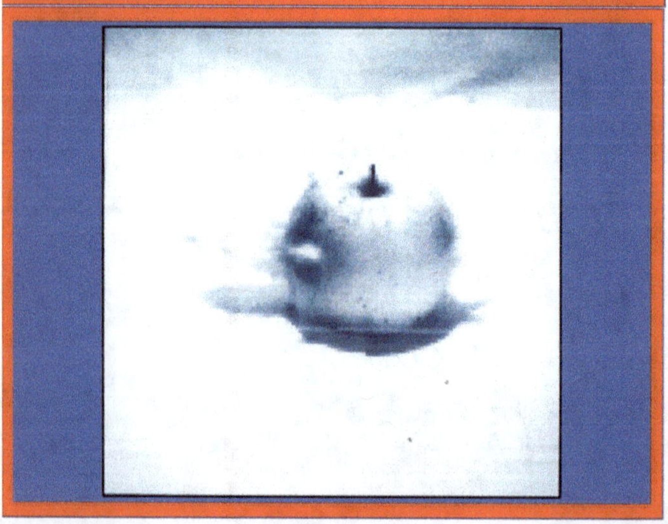

1-84

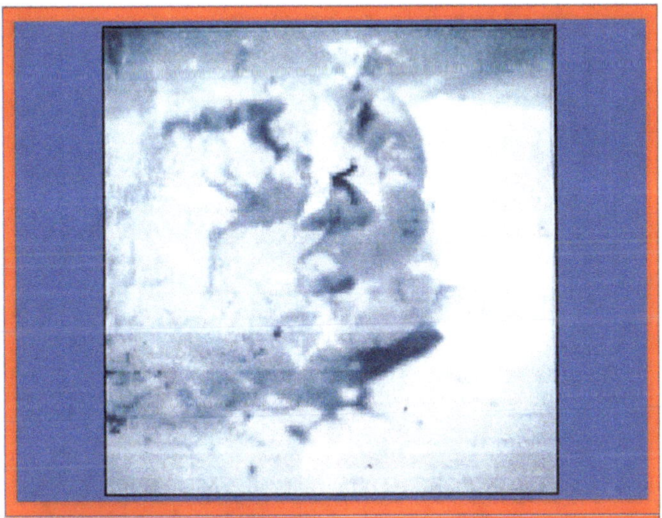

1-85

1-86

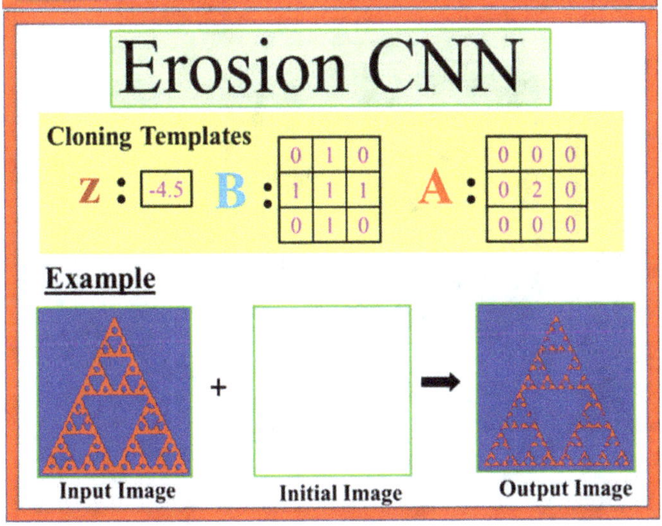
Erosion CNN
Cloning Templates
z : -4.5
B : 0 1 0 / 1 1 1 / 0 1 0
A : 0 0 0 / 0 2 0 / 0 0 0
Example
Input Image
+
Initial Image
Output Image

1-87

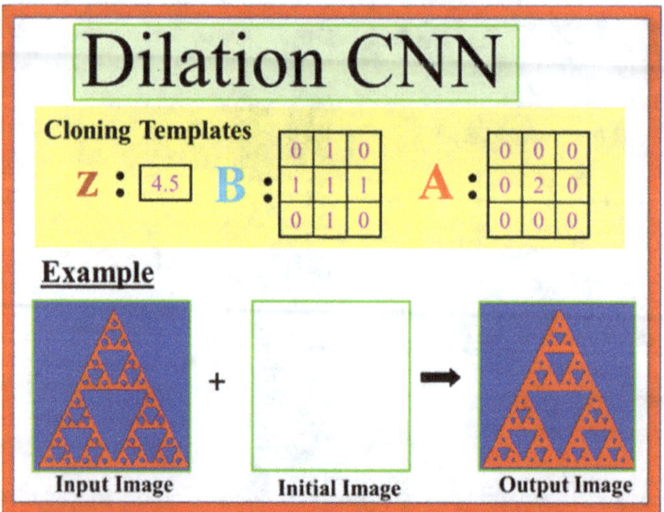
Dilation CNN
Cloning Templates
z : 4.5
B : 0 1 0 / 1 1 1 / 0 1 0
A : 0 0 0 / 0 2 0 / 0 0 0
Example
Input Image
+
Initial Image
Output Image

1-88

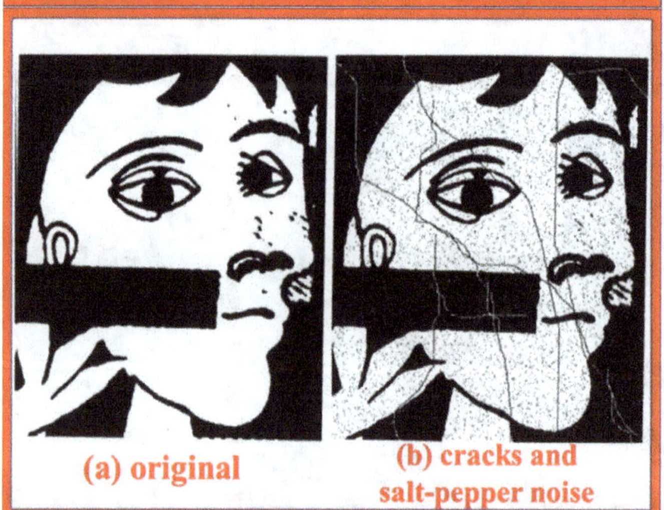
(a) original
(b) cracks and salt-pepper noise

1-89

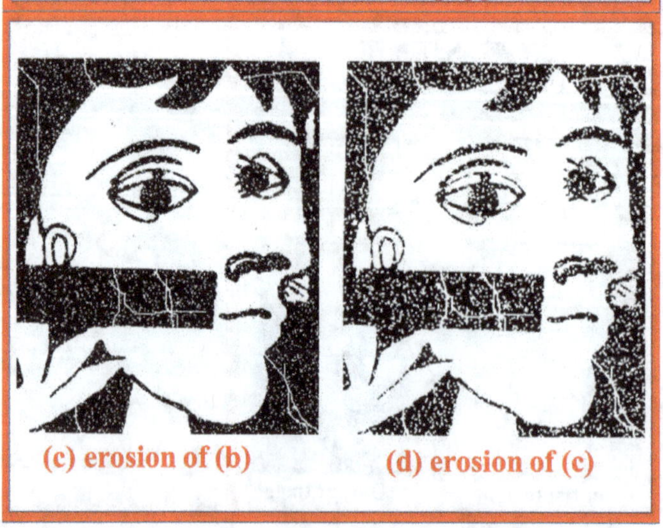
(c) erosion of (b)
(d) erosion of (c)

1-90

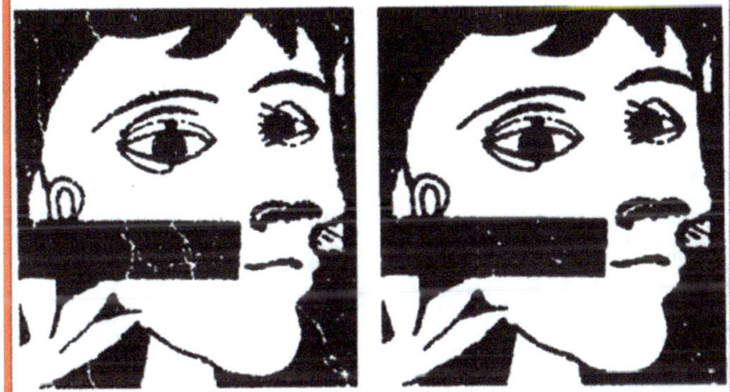

(e) dilation of (d) (f) dilation of (e)

1-91

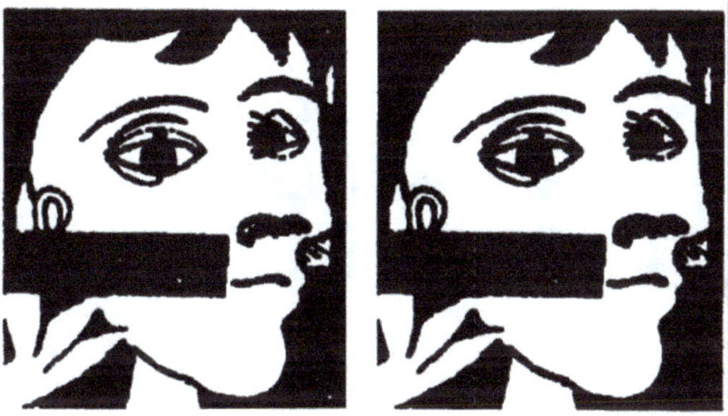

(g) dilation of (f) (h) dilation of (g)

1-92

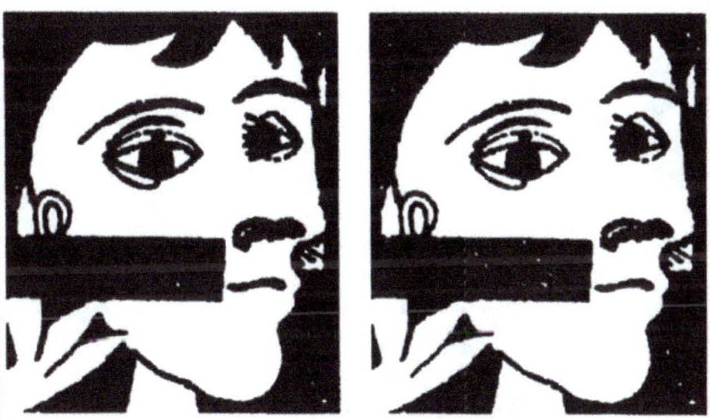

(i) erosion of (h) (j) erosion of (i)

1-93

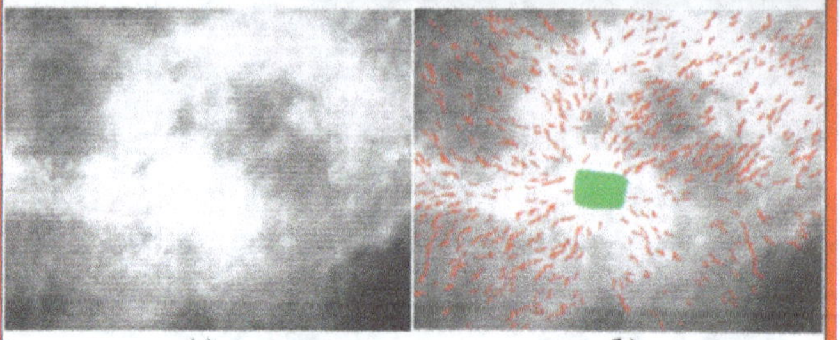

(a) (b)

*The short red radial threads are **spiculi** detected by a CNN program from a **mammogram**,*

confirming a cancer tumor.

1-94

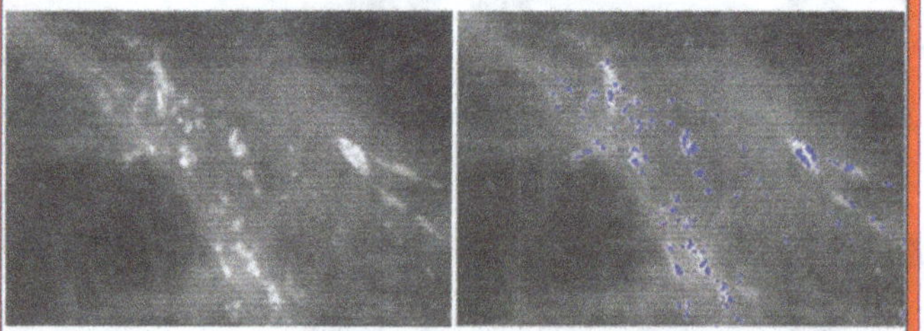

(a) (b)

*The blue particles are **micro-calcifications** detected by a CNN program from a **mammogram**,*

confirming a cancer tumor.

1-95

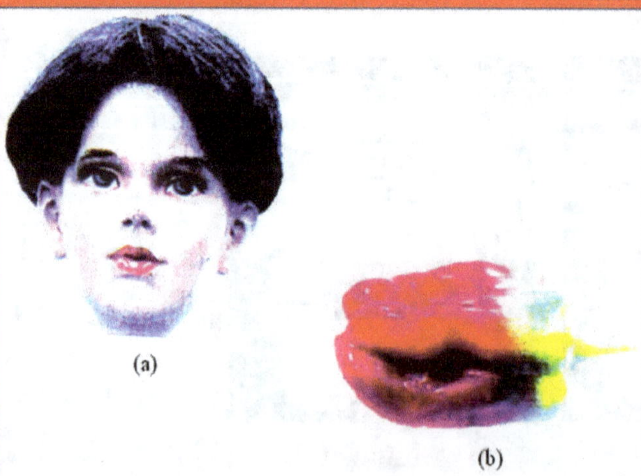

(a) (b)

1-96

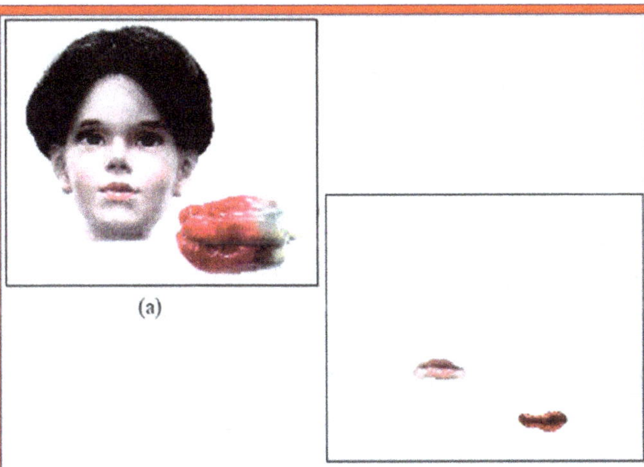

(a) (b)

1-97

Part 3 Local Activity and Edge of Chaos

1-98

Hodgkin-Huxley Spike Mystery

How does the neuron generate action Potentials (spikes) ?

1-99

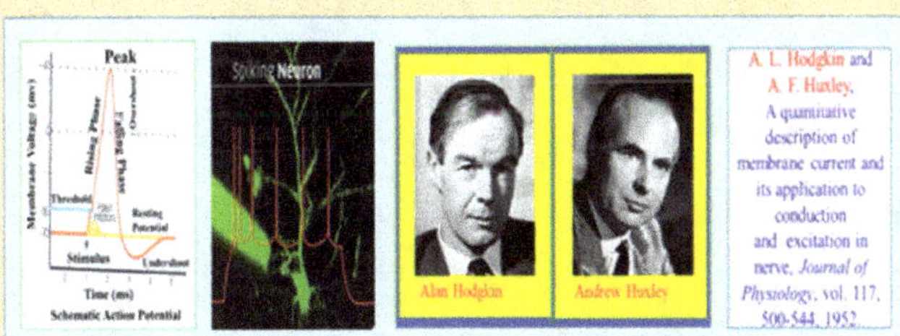

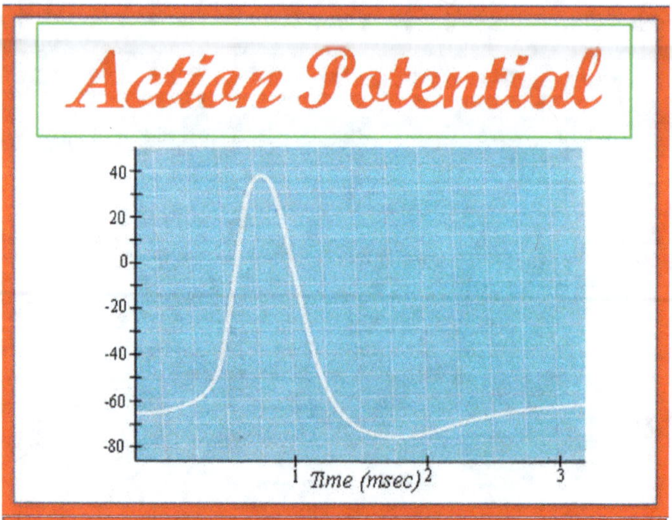

1-100

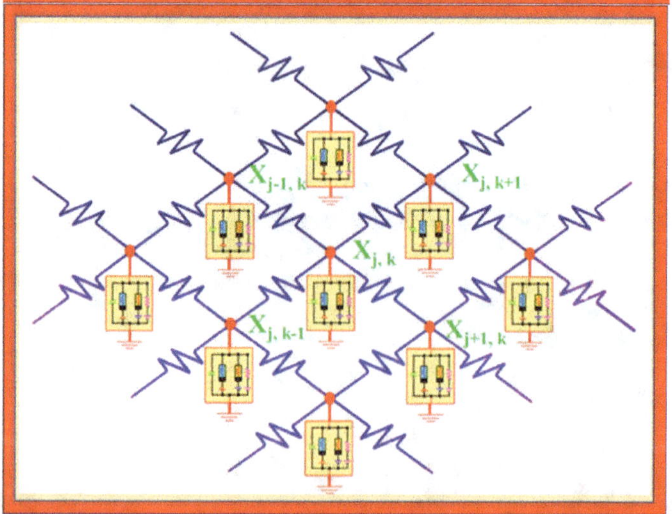

1-101

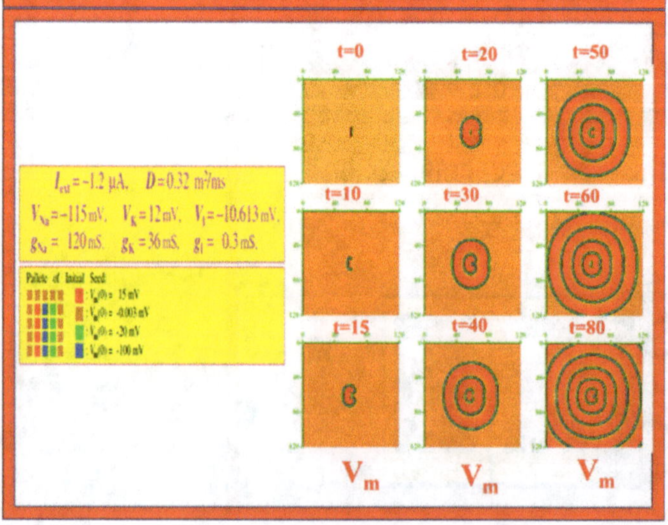

1-102

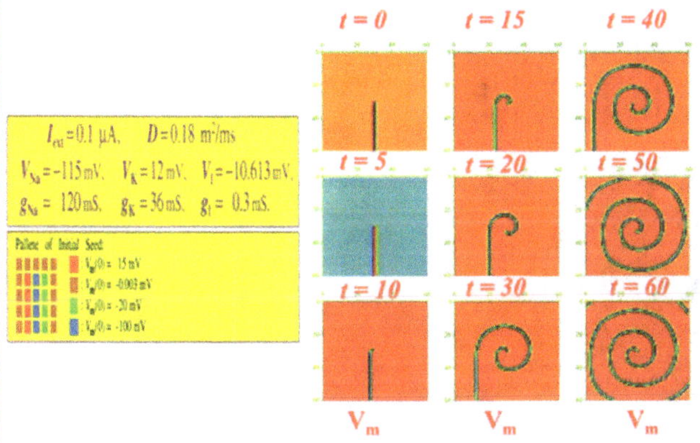

1-103

Spiral Ca^{2+} wave propagated in an Oocyte.

1-104

The Unknown

Mechanism

of the abrupt

Hodgkin-Huxley's

Action Potential

is the

Edge of Chaos

1-105

1-106

Three Fundamental Concepts

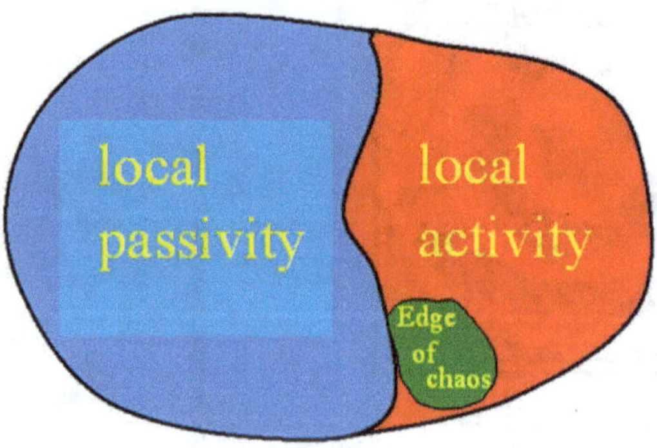

1-107

The Principle of Local Activity

pertains to

OPEN SYSTEMS

where entropy is a non-monotonic function

It is the missing

Complementary Principle

of the

Second Law of Thermodynamics

1-108

The 4th Law of Thermodynamics

Complexity is impossible without Local Activity *1*-109

Edge of Chaos is the "pearl" Imbedded within the domain of local activity *1*-110

Neurons are poised near the Edge of Chaos *1*-111

Our

Brain

can make

Spiral waves and

Target waves ! 1-112

If the range of

Edge of Chaos

in the

Hodgkin-Huxley Neuron

were to be 1-113

"UNTUNED"

by more than 2 percent

there would be

no intelligent life

Life

is

Impossible 1-114

without

Edge of Chaos

The

Mechanism

of

Edge of Chaos *1*-115

is

The Chua Riddle

Chua's

Riddle *1*-116

Chua's Riddle

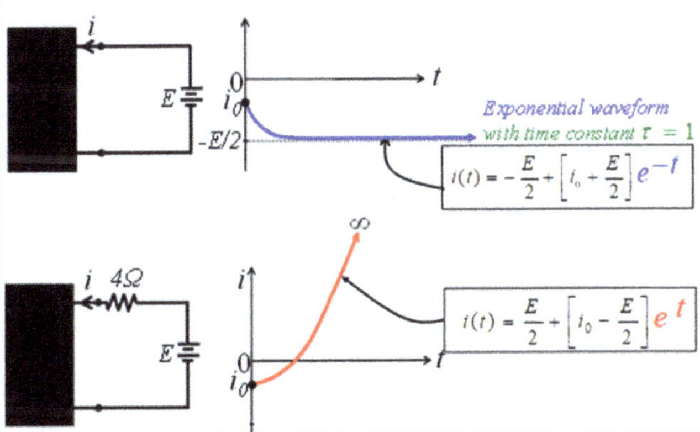

1-117

What is inside the Black Box  1-118

Chua's Riddle 1-119

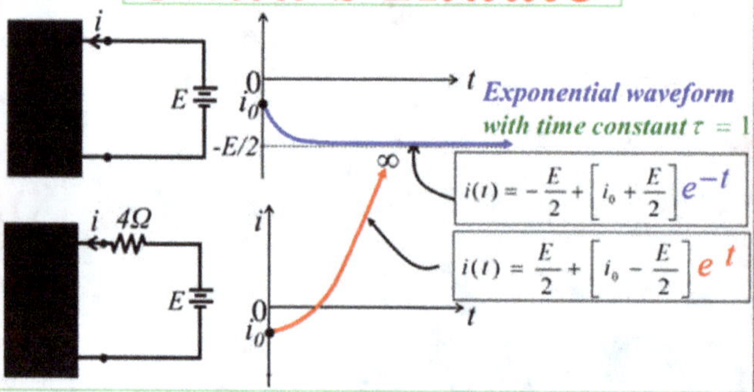

Hint : Black Box contains 2 linear basic circuit elements.

One-Dimensional CNN 1-120

is a simple yet powerful theory for studying

Complexity

and

Universal Turing Machines

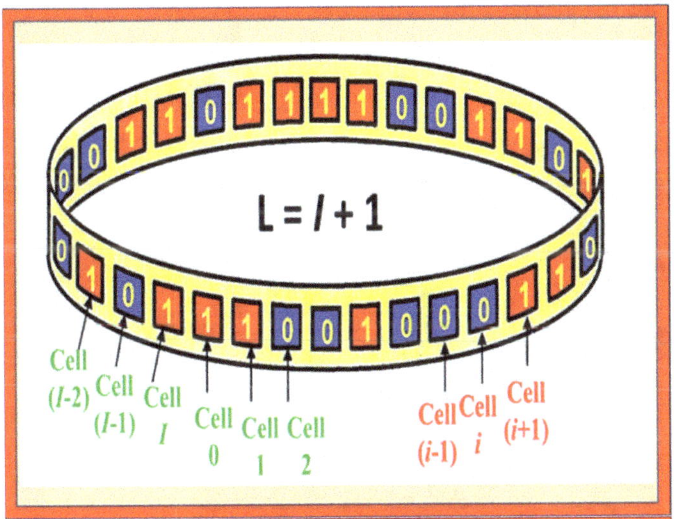

I-121

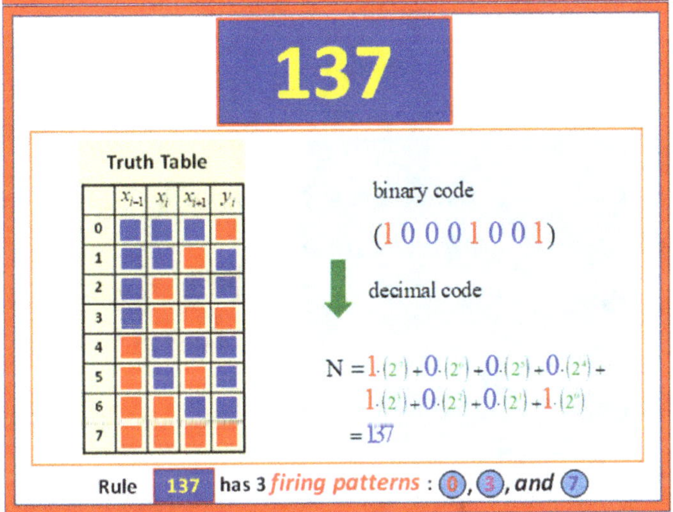

I-122

CNN can mimic 137

137 can mimic

A Turing Machine

Therefore,

CNN can mimic

A Turing Machine

I-123

hp has demonstrated a *memristor circuit* model which executes a *Universal Turing Machine* *via* the *Cellular Automata* **Rule 137** 1-124

Climax *of* *The Chua Lectures*

The foundation and applications of *Wolfram's Monumental book* 1-125

A new kind of Science

will be presented and illustrated via metaphors from Physics, Quantum Mechanics, and Social Sciences.

1-126

End Quote for Lecture 1

To know that one knows what one knows, and to know that one doesn't know what one doesn't know, there lies true wisdom.

CONFUCIUS

1-127

Appendix

1-128

Discussion

1-129

Lecture 2

Everything You Wish to Know About Memristors but are Afraid to Ask

1-130

Lecture 3

10 Things You Didn't Know about Memristors

1-131

Videos

1-132

Frequently Asked Question

Why did it take so long for your memristor theory to be validated ?

1-133

New Scientific ideas do not succeed by converting contemporary scientists, but rather by their opponents' dying off

Max Planck

Max Planck

1-134

New theories have four stages of acceptances:

I. this is worthless nonsense;

J. B. S. Haldane

J. B. S. Haldane

1-135

New theories have four stages of acceptances:

I. this is worthless nonsense;
II. this is interesting, but perverse;

J. B. S. Haldane

J. B. S. Haldane

1-136

New theories have four stages of acceptances:

I. this is worthless nonsense;
II. this is interesting, but perverse;
III. this is true, but quite unimportant;

J. B. S. Haldane

J. B. S. Haldane

1-137

New theories have four stages of acceptances:

I. this is worthless nonsense;
II. this is interesting, but perverse;
III. this is true, but quite unimportant;
IV. I always said so.

J. B. S. Haldane

J. B. S. Haldane

1-138

1-139

1-140

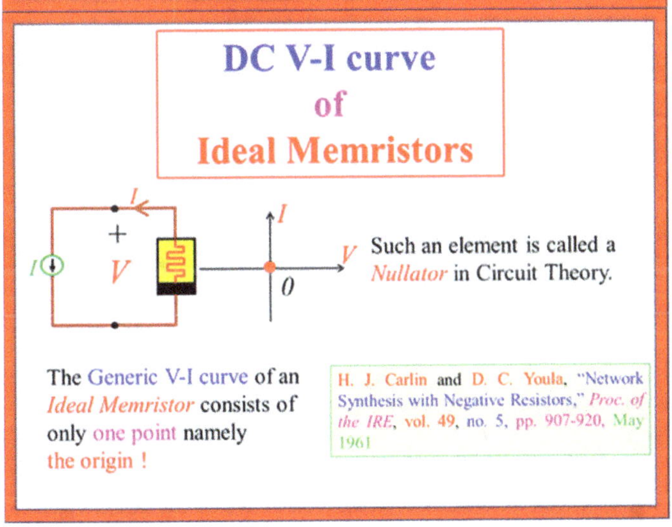

1-141

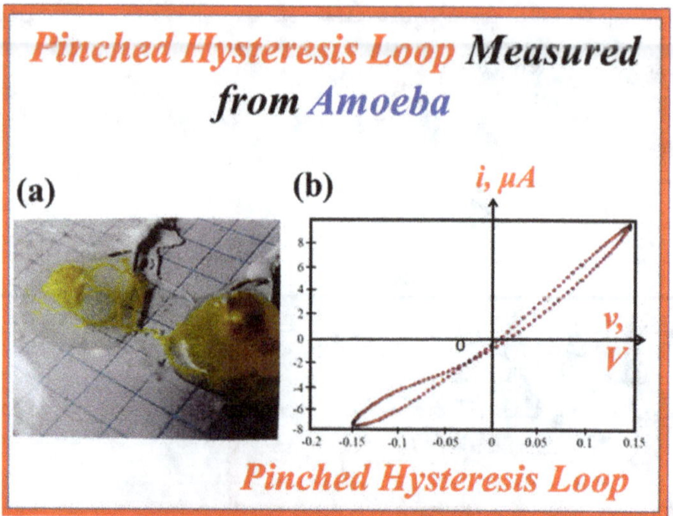

1-142

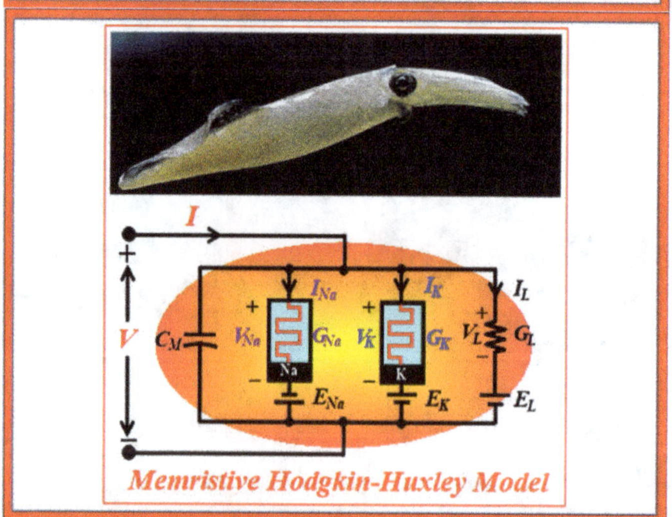

1-143

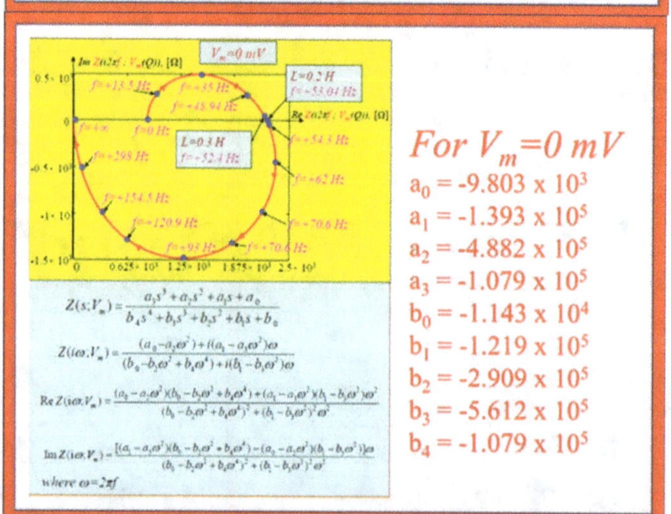

1-144

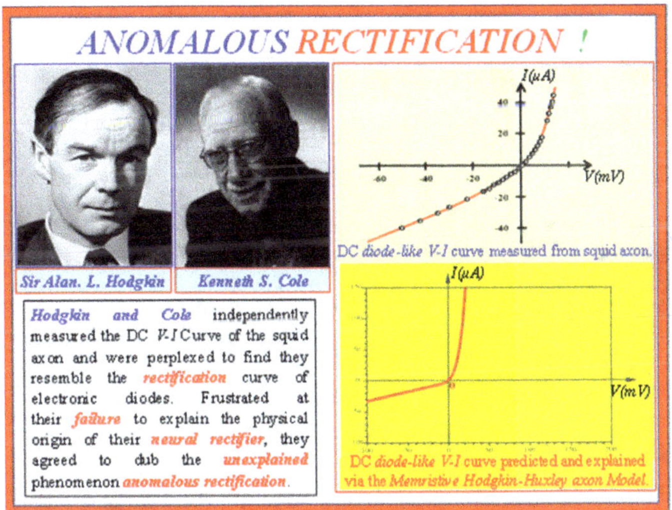

1-145

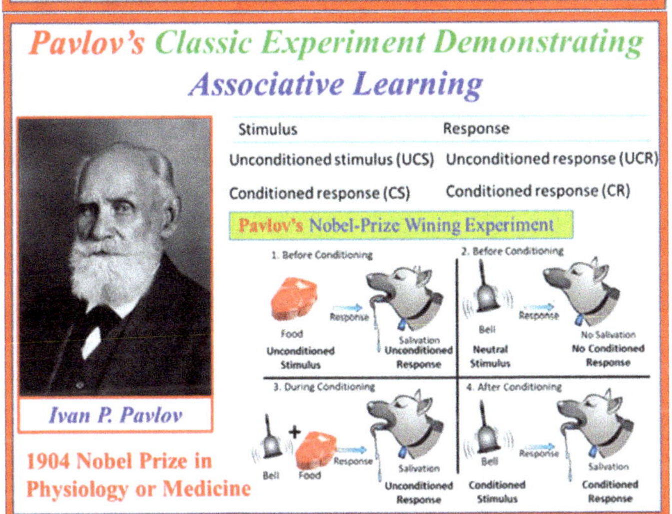

1-146

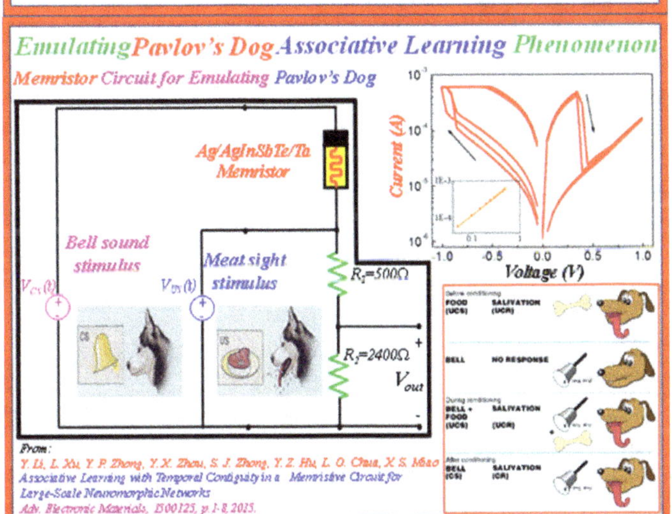

1-147

CNN:

A Brain-Like Supercomputer on A Chip 1-148

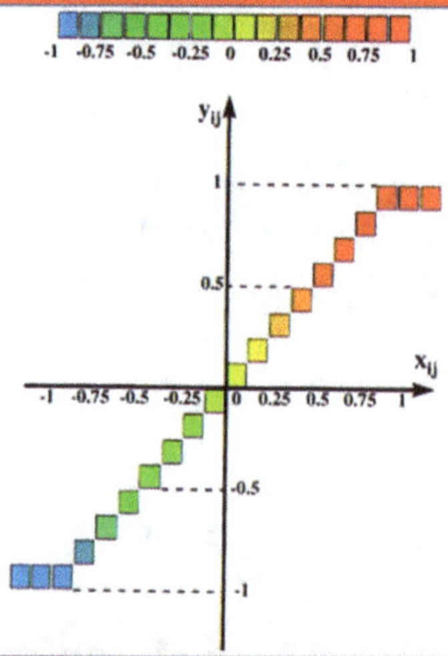

1-149

CNN Flows via Nonlinear Dynamics

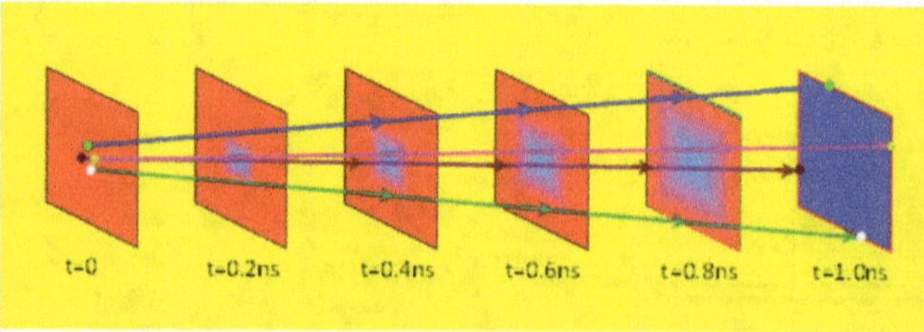

1-150

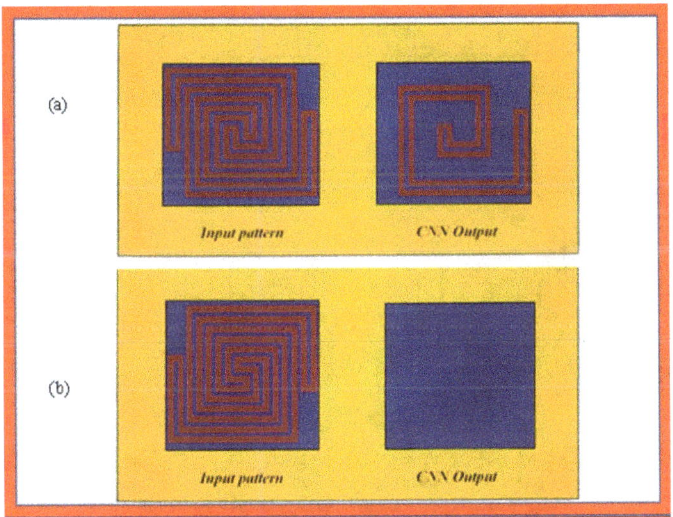

1-151

J. M. Cruz and L. O. Chua
"A 16 x 16 cellular neural network universal chip: The first complete single-chip dynamic computer array with distributed memory and with gray-scale input-output",
Analog Integrated Circuits Signal Processing Journal
vol. 15, no. 3, pp. 227-238, 1998.

1-152

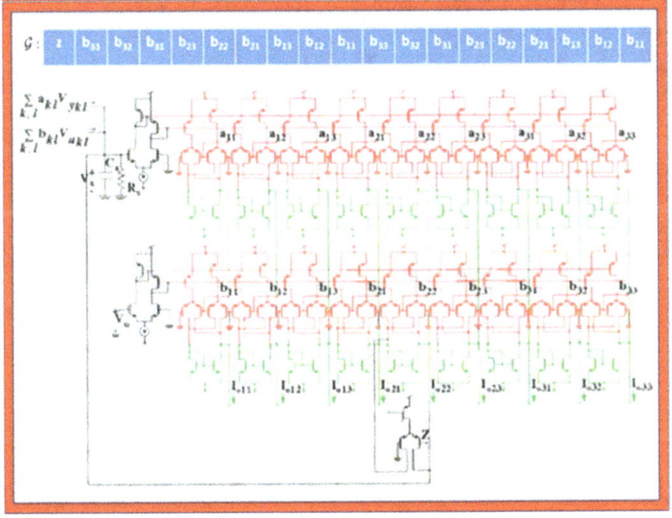

1-153

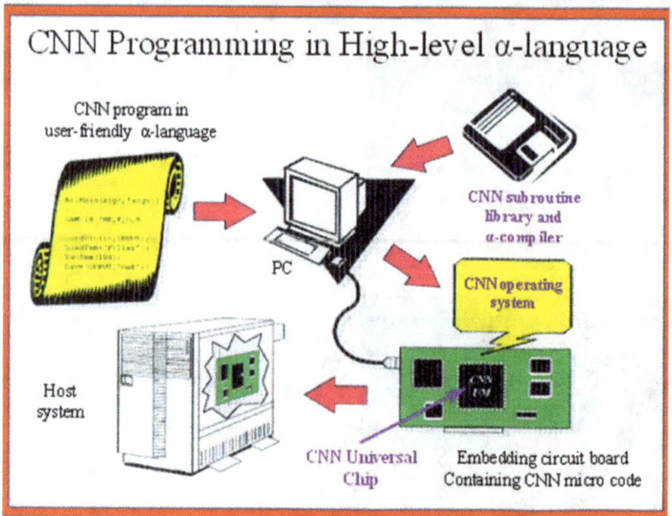

1-154

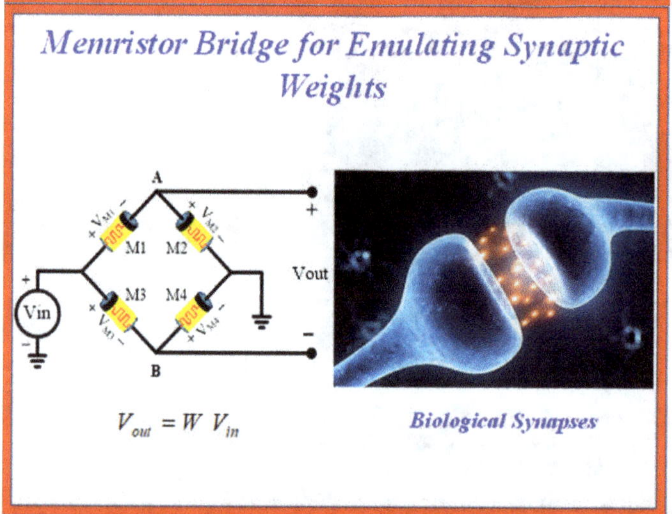

1-155

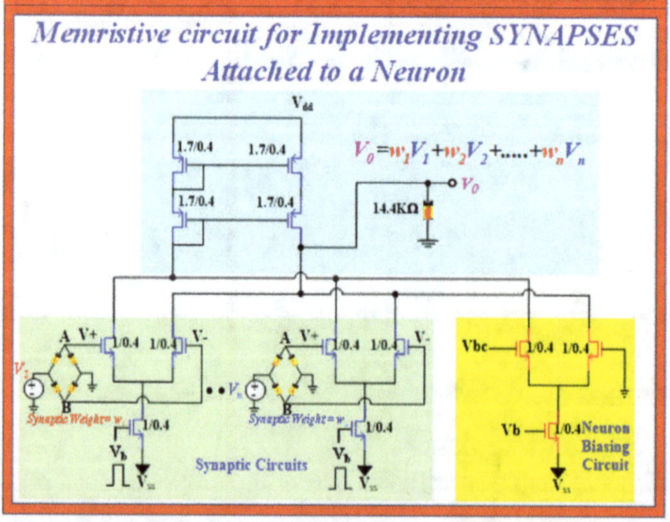

1-156

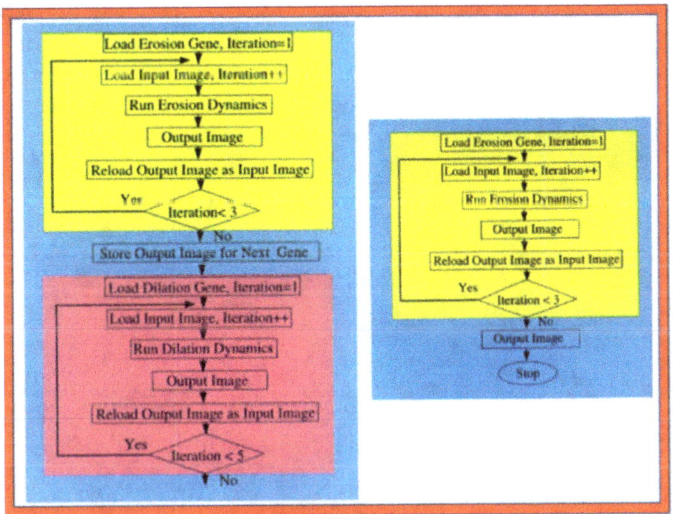

1-157

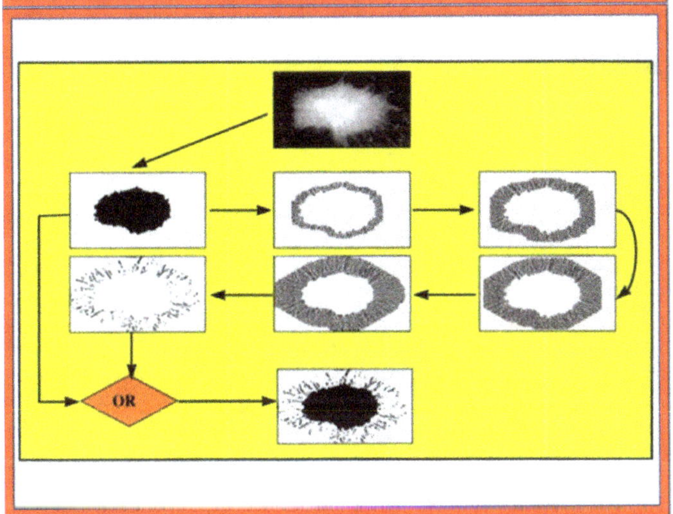

1-158

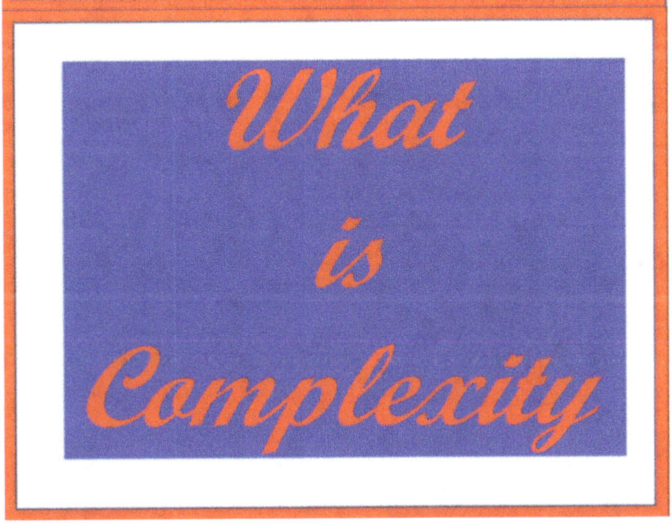

1-159

What is Not Complexity? *1*-160

1-161

1-162

The evolution towards a

HOMOGENEOUS

STEADY STATE

is the Hallmark of

The Second Law of Thermodynamics

1-163

The Second Law of Thermodynamics

predicts a boring

Lifeless and Featureless

scenario for

Isolated Systems

1-164

How does the ***Leopard*** *get its* **spots** *?*

1-165

How does the **Leopard's tail** get its **bands ?** 1-166

How does the **Tiger** get its **Stripes ?** 1-167

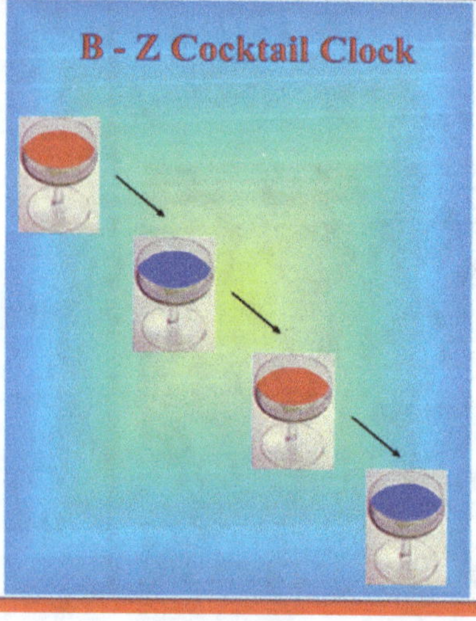

1-168

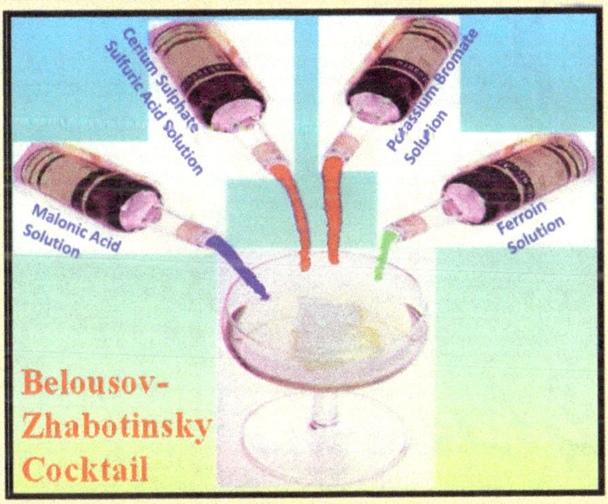

1-169

The spectacular patterns of the Belousov-Zhabotinsky reaction consist of target and spiral waves.

1-170

Definition : Complexity

Complexity

is defined as the emergence of a ***heterogeneous pattern*** *on a* ***homogeneous media.***

1-171

Pattern is Information

1-172

Ludwig Boltzmann

Boltzmann's Characterization of Complexity :

Entropy must decrease

1-173

Erwin Schrödinger

Schrödinger's Characterization of Complexity :

Negentropy

1-174

Ilya Prigogine

Prigogine's
Characterization
of Complexity :

Instability of
the Homogeneous

1-175

Murray Gell-Mann

Gell Mann's
Characterization
of Complexity :

Amplification of
Fluctuations

1-176

Alan M. Turing

Turing's
Characterization
of Complexity:
Destabilize
by
Dissipation

1-177

Schrödinger, Prigogine,
Eigen, Gell-Mann, Turing,
Smale, etc.
have all been searching for a
missing new
Physical Principle !

1-178

The
Missing New Principle
is called
Local Activity
and its associated
Edge of Chaos

1-179

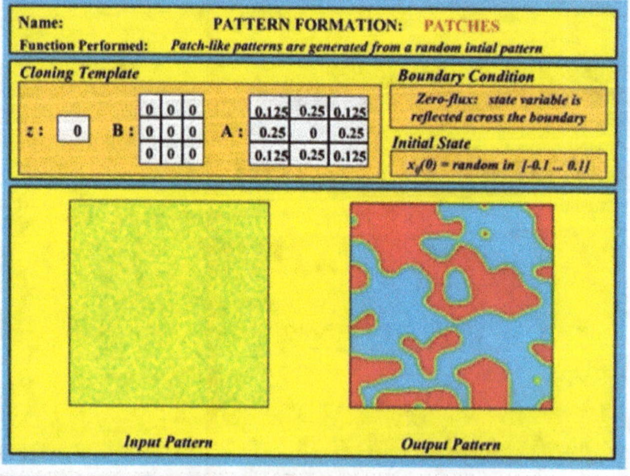

1-180

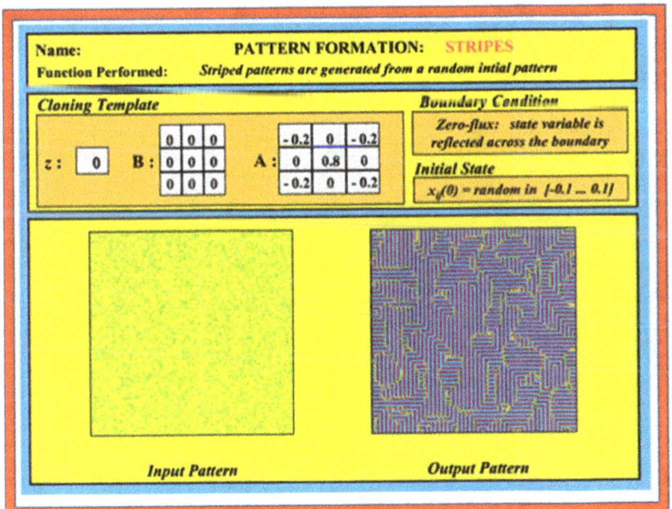

1-181

CNN Templates for Emulating Tiger Stripes and Leopard Spots

z: $0 < z < 1.0$

B:

0	0	0	0	0
0	0	0	0	0
0	0	0	0	0
0	0	0	0	0
0	0	0	0	0

A:

-0.25	-1.0	-1.5	-1.0	-0.25
-1.0	2.5	7.0	2.5	-1.0
-1.5	7.0	-23.25	7.0	-1.5
-1.0	2.5	7.0	2.5	-1.0
-0.25	-1.0	-1.5	-1.0	-0.25

1-182

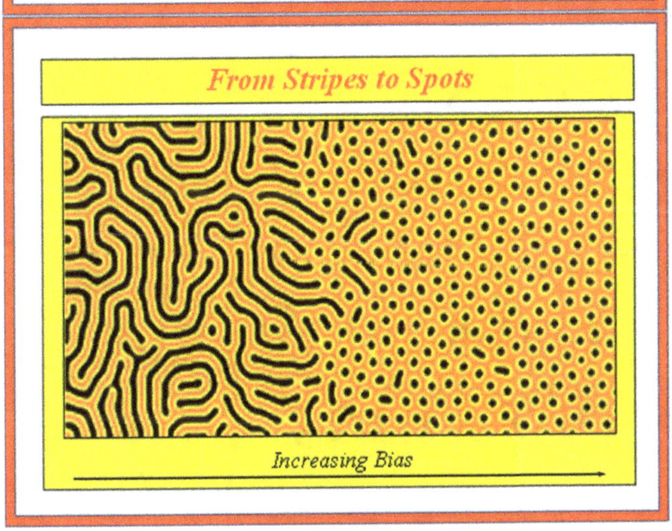

1-183

CNN Leopard Tail

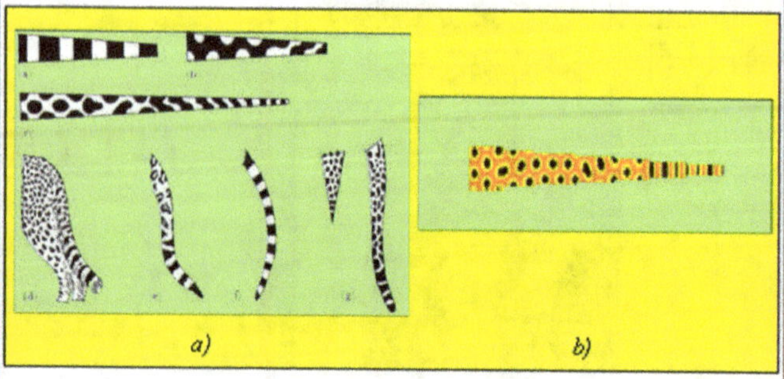

1-184

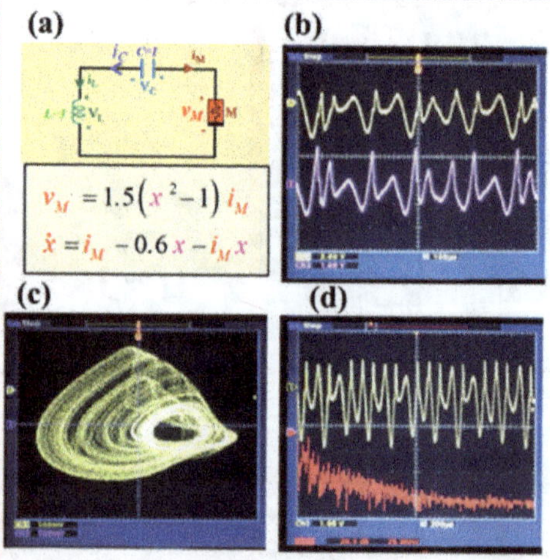

The simplest possible chaotic circuit where the equation describes a Generic Memristor with memristance *R(x) = 1.5 (x²-1)* 1-185

Turing's instability via dissipation phenomenon originates from the Edge of Chaos 1-186

Good News !

Edge of Chaos is Testable !

Testing for
Edge of Chaos
involves only
Linear Algebra

1-187

Fundamental New Result

Edge of Chaos Theorem

A stable equilibrium state Q can be destabilized by coupling to a dissipative environment if, and only if, Q is poised on, or near, the

Edge of Chaos

1-188

Proof of the Edge of Chaos Theorem requires non-trivial applications of the following mathematical results:

1. Nonlinear Monotone Operator in Function Space
2. High-dimensional Analog of the Mean-Value Theorem
3. LaSalle's Complete Stability Theorem
4. Poisson's Integral on half-plane
5. Theory of Positive-Real Function of a complex Variable
6. Tellegen's Theorem

1-189

Riddle

Tamed when open

Tamed when shorted

Explodes when dissipated 1-190

Chua's Riddle

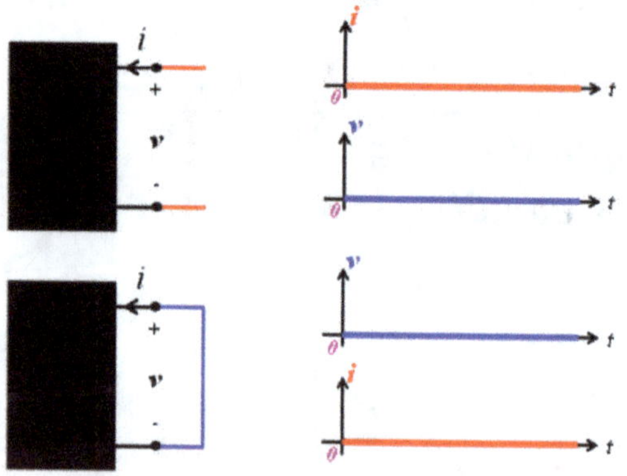

1-191

Memristors alone (without using transistors) can, in principle, be used to build any system that can solve any problem solvable by a Supercomputer. 1-192

1-193

References on Memristors

1. L. O. Chua Memristor – The Missing Circuit Element IEEE Trans. on Circuit Theory, vol. 18, pp. 507-519, 1971.
2. L. O. Chua, S. M. Kang Memristive Devices and Systems Proceedings of IEEE, vol. 64, pp. 209-223, 1976.
3. D. B. Strukov, G. S. Snider, D. R. Stewart, R. S. Williams The Missing Memristor Found Nature 453, pp.80-83, 2008.
4. R. S. Williams How we Found the Missing Memristor IEEE Spectrum, 2008.
5. L Chua Introduction to Memristors IEEE Expert Now Short Course,2009 http://ieeexplore.ieee.org/courses/details/EDP091.

1-194

6. L. Chua Resistance Switching Memories Are Memristors Applied Physics A, 102, pp. 765-783, 2011.
7. L. Chua The Fourth Element Proceedings of IEEE, vol. 100, pp.1920-1927, 2012
8. L. Chua Memristor, Hodgkin-Huxley, and Edge of Chaos Nanotechnology, 24, pp.383001(14pp), 2013.
9. M. P. Sah, H. Kim, L. Chua Brains Are Made of Memristors IEEE Circuits and Systems Magazine, vol.14, pp.12-36, 2014.
10. L. Chua If It's Pinched It's a Memristor Semicond. Sci. Technol., 29, pp.104001(42pp), 2014.
11. L. Chua Everything You Wish to Know About Memristor But Are Afraid to Ask RADIOENGINEERING, vol.24, pp.319-368, 2015.
12. L. Chua, How we predicted the memtistor Nature Electronics, vol. 1, 2028, p.32.

1-195

References on Cellular Neural Networks

1. L. O. Chua and L. Yang
   Cellular Neural Networks: Theory and Applications
   IEEE Trans. on Circuits and Systems, vol.35, pp.1257-1290, 1988.

2. L. O. Chua
   CNN: A Paradigm for Complexity
   World Scientific, Singapore, 1998.

3. J. M. Curz and L. O. Chua
   A 16x16 cellular neural network universal chip: The first complete single – chip dynamic computer array with distributed memory and with gray – scale input-output.
   Analog Integrated Circuits Signal Processing Journal, vol.15, no.3, pp.227-238, 1998 .

4. L. O. Chua and T. Roska
   Cellular Neural Networks and Visual Computing
   Cambridge University Press, 2002

1-196

References on Local Activity and Edge of Chaos

1. L. O. Chua
   CNN: A Paradigm for Complexity
   World Scientific, Singapore, 1998

2. L. O. Chua
   Passivity and Complexity
   IEEE Trans. On Circuits and Systems, vol. 46, no.01, 1999.

3. R. Dogaru, L. O. Chua
   Edge of Chaos and Local Activity Domain of FitzHugh-Nagumo Equation
   International Journal of Bifurcation and Chaos, vol.08, no. 02, 1988.

4. L. O. Chua
   Local Activity is the Origin of Complexity
   International Journal of Bifurcation and Chaos, vol.15, no. 11, 2005.

1-197

References on Local Activity and Edge of Chaos

5. L. Chua
   Memristor, Hodgkin-Huxley, and Edge of Chaos
   Nanotechnology, 24, pp.383001(14pp), 2013.

6. L. Chua, V. Sbitnev, H. Kim
   Hodgkin–Huxley Axon is Made of Memristors
   International Journal of Bifurcation and Chaos, vol.22, no.03, 2012.

7. L. Chua, V. Sbitnev, H. Kim
   Neurons Are Poised Near the Edge of Chaos
   International Journal of Bifurcation and Chaos, vol.22, no.04, 2012.

8. K. Mainzer, L. Chua
   Local Activity Principle: The Cause of Complexity and Symmetry Breaking
   Imperial College Press, 2013.

1-198

Lecture 2

Everything You Wish to Know About Memristors but are Afraid to Ask

Lecture 2

Everything You Wish to Know About Memristors but are Afraid to Ask

2-1

1960-1970

The Electrical Engineers

Cambrian Explosion

2-2

1960

CRISIS

IN

Electronic Circuit Theory

2-3

Esaki Diode
I
i_1
$r_{diff} < 0$
i_2
V_1
V_2
V
Leo Esaki
Nobel Prize in Physics, 1973
L. Esaki
New Phenomenon in Narrow Germanium p-n junctions
Physics review 109(2):603, 1958

2-4

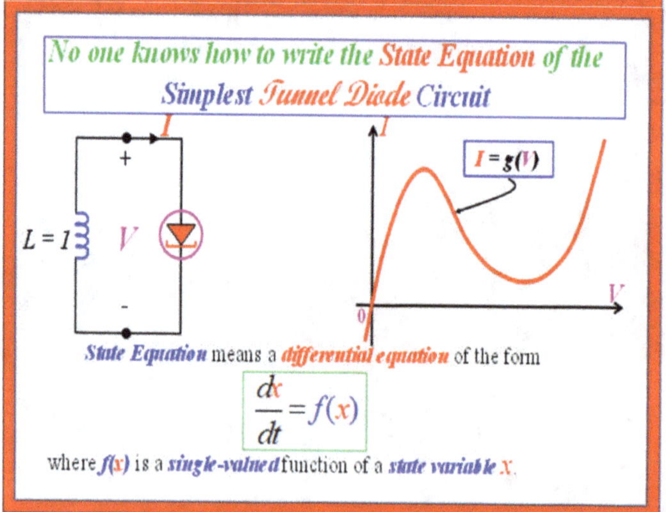
No one knows how to write the State Equation of the Simplest Tunnel Diode Circuit
I
+
L = 1
V
-
I
I = g(V)
V
0
State Equation means a differential equation of the form
$\frac{dx}{dt} = f(x)$
where f(x) is a single-valued function of a state variable x.

2-5

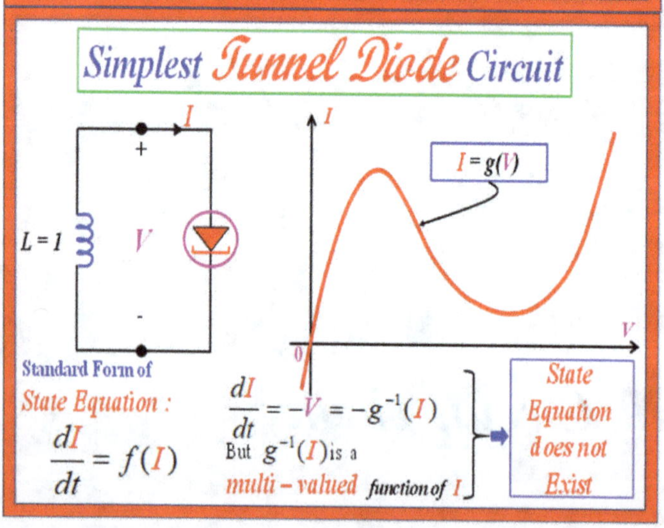
Simplest Tunnel Diode Circuit
I
+
L = 1
V
-
I
I = g(V)
V
0
Standard Form of
State Equation :
$\frac{dI}{dt} = f(I)$
$\frac{dI}{dt} = -V = -g^{-1}(I)$
But $g^{-1}(I)$ is a
multi – valued function of I
State Equation does not Exist

2-6

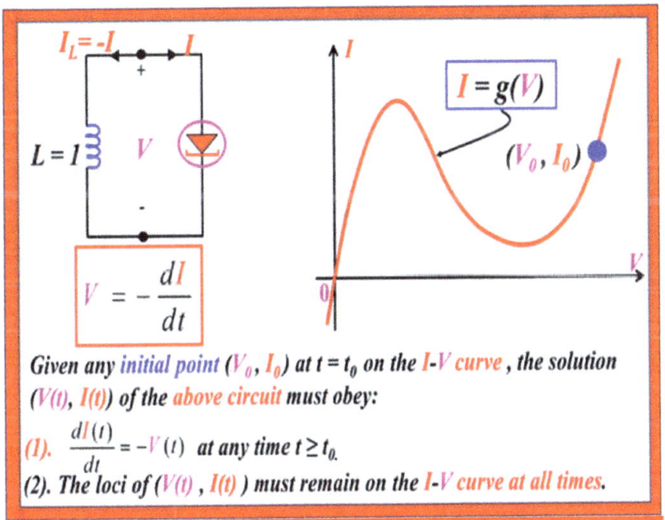

I_L = -I
I
L = 1
V
+
-
V = -dI/dt
I
I = g(V)
(V_0, I_0)
0
V
Given any initial point (V_0, I_0) at t = t_0 on the I-V curve, the solution (V(t), I(t)) of the above circuit must obey:
(1). dI(t)/dt = -V(t) at any time t ≥ t_0.
(2). The loci of (V(t), I(t)) must remain on the I-V curve at all times.

2-7

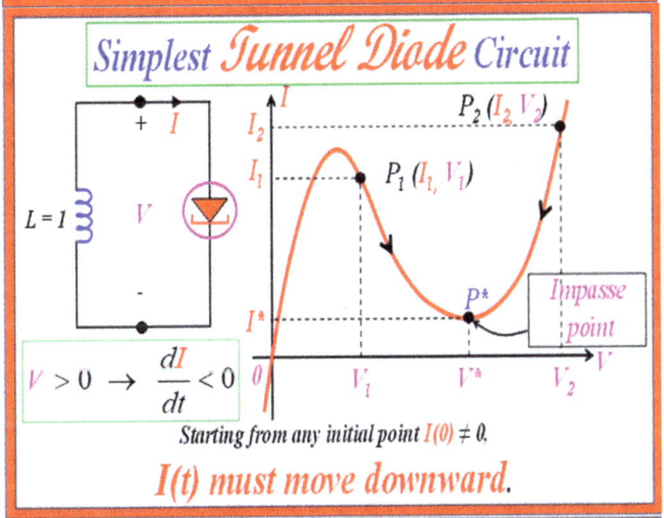

Simplest Tunnel Diode Circuit
+
I
L = 1
V
-
V > 0 → dI/dt < 0
I
I_2
I_1
I*
0
P_2 (I_2, V_2)
P_1 (I_1, V_1)
P*
Impasse point
V_1
V*
V_2
V
Starting from any initial point I(0) ≠ 0,
I(t) must move downward.

2-8

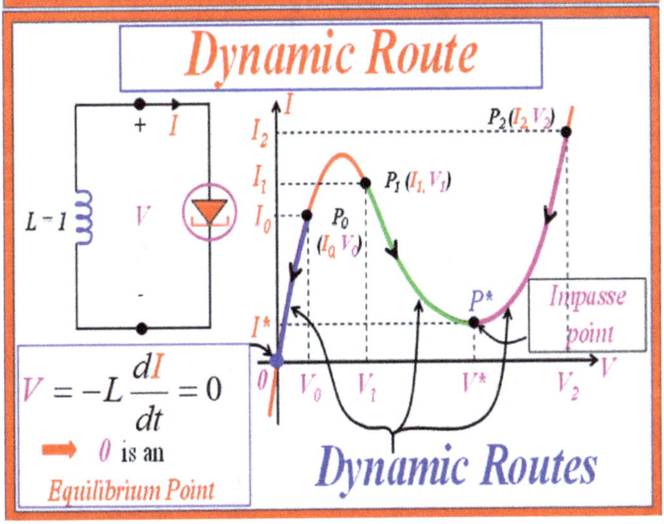

Dynamic Route
+
I
L = 1
V
-
V = -L dI/dt = 0
0 is an Equilibrium Point
I
I_2
I_1
I_0
I*
0
P_2 (I_2, V_2)
P_1 (I_1, V_1)
P_0 (I_0, V_0)
P*
Impasse point
V_0
V_1
V*
V_2
V
Dynamic Routes

2-9

Simplest *Tunnel Diode* Circuit

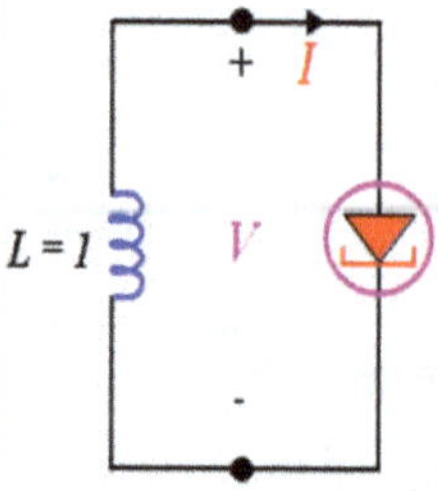

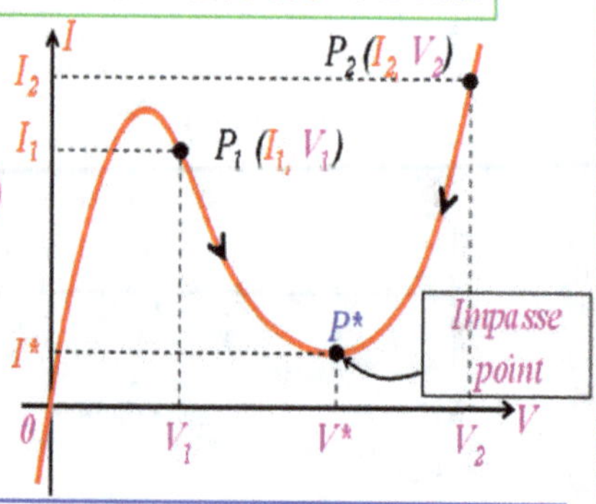

$$V = -\frac{dI}{dt}$$ 2-10

Solution does NOT exist beyond P* !

Two Points of View !

- For *mathematician*, *no solution* is a Perfectly *valid solution* 2-11
- For *everybody else*, no solution means *nonsence*.

To write state Equation We must Add A Parasitic Capacitor

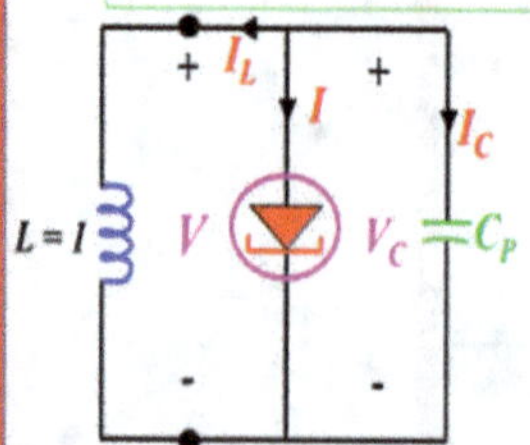

State Equations :

$$\frac{dI_L}{dt} = -V_C$$

$$\frac{dV_C}{dt} = \frac{1}{C_P}[-I_L - g(V_C)]$$ 2-12

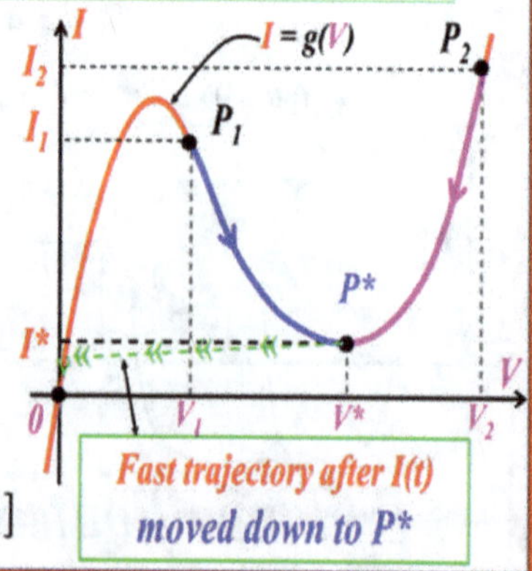

Take Home Lesson 1

The best model
is the simplest model
that gives a
realistic solution. 2-13

Bad Habit to Unlearn

Linearize,
then
analyze 2-14

How does Oscillator Work ?

The linearized oscillation
mechanism you learned from
standard textbooks is
wrong ! 2-15

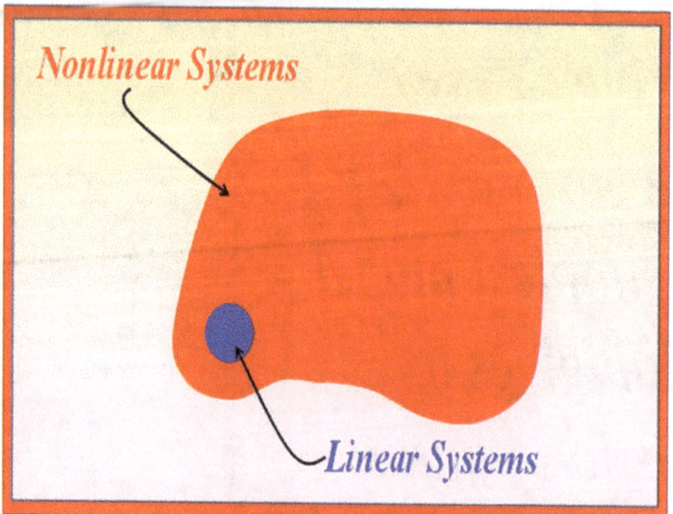
Nonlinear Systems
Linear Systems

2-16

The Blind Men and The Elephant
Each man experiences something different depending on which part of the elephant he touches.

2-17

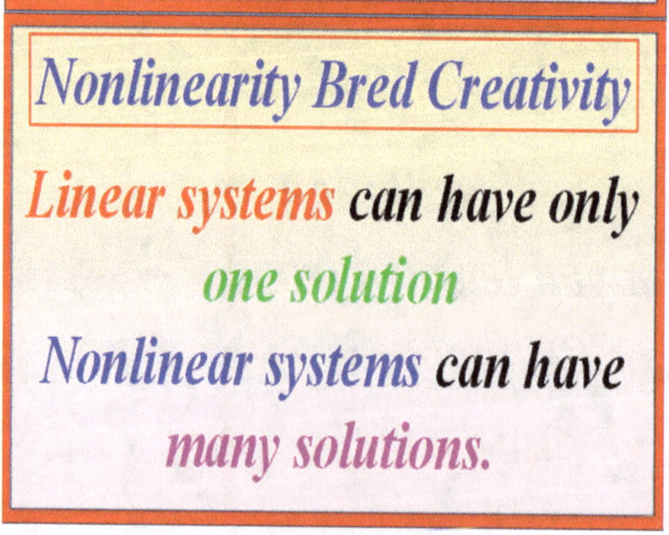
Nonlinearity Bred Creativity
Linear systems can have only one solution
Nonlinear systems can have many solutions.

2-18

Take Home Lesson 2

Don't Linearize!

2-19

Varactor Diode

Developed by Ramo Wooldridge Corporation patent issued 1961

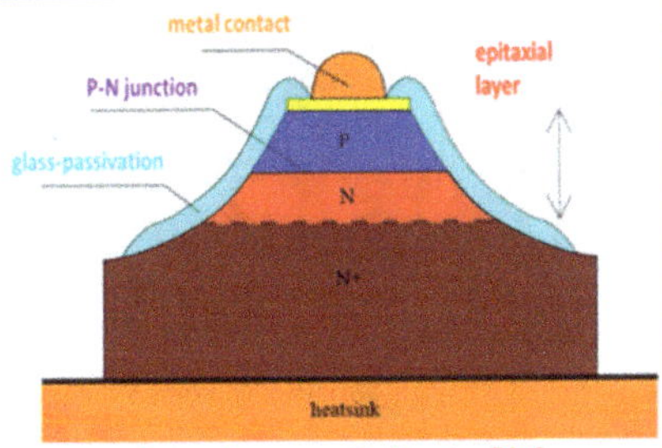

Internal structure of a Varactor Diode

Type	passive
Pin configuration	anode and cathode

Electronic symbol

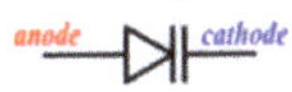

2-20

3 Basic Circuit Elements

1745		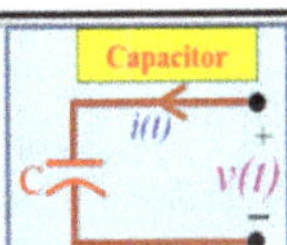	$i(t) = C\dfrac{dv(t)}{dt}$
1827		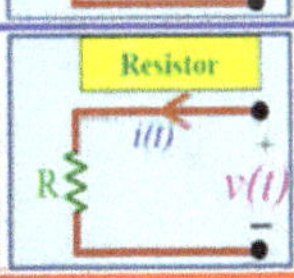	$v(t) = R\,i(t)$
1831		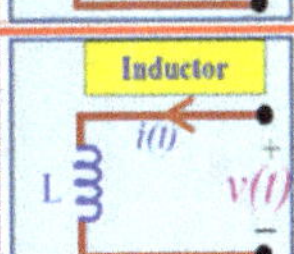	$v(t) = L\dfrac{di(t)}{dt}$

2-21

Example : Time-varying Capacitance

$$C(t) = 10 + 5\sin t$$

Wrong Formula !

$$i(t) = C(t)\,\frac{dv(t)}{dt}$$ 2-22

$$\Rightarrow\ i(t) = (10 + 5\sin t)\,\frac{dv(t)}{dt}$$

Correct formula:

$$q(t) = C(t)\,v(t)$$

$$i = \frac{dq(t)}{dt} = C(t)\frac{dv(t)}{dt} + \underbrace{v(t)\frac{dC(t)}{dt}}_{\text{extra term is needed !}}$$

$$q(t) = \underbrace{(10 + 5\sin t)}_{C(t)}\,v(t)$$ 2-23

$$i = \frac{dq(t)}{dt} = \underbrace{(10 + 5\sin t)}_{C(t)}\frac{dv(t)}{dt} + (5\cos t)\,v(t)$$

Conclusion:
A *time-varying capacitance* $C(t)$ must be defined by a relationship between the *charge* $q(t)$ and the *voltage* $v(t)$, and *not* between $i(t)$ and $\frac{dv(t)}{dt}$, as in the *incorrect* formula $i = C(t)\frac{dv}{dt}$

Rocket Launching

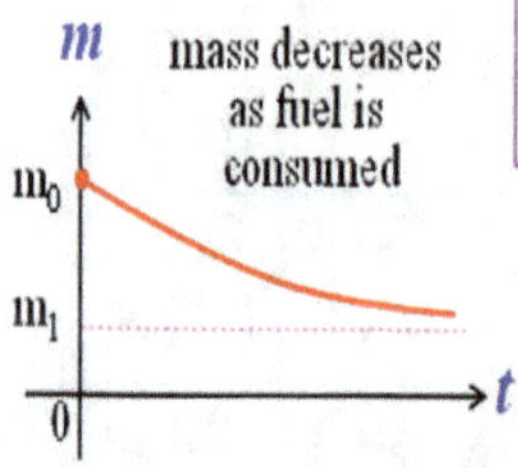

Newton's Law of Motion 2-24

$$f(t) = m(t)\frac{dv}{dt}$$

Is this formula correct ?

NO !
Correct Newton's Formula is:

$$f = \frac{dp}{dt}$$

$$p = mv \quad \leftarrow \text{momentum}$$

For *time-varying mass*, we have :

$$p = m(t)\, v$$

∴

$$f(t) = m(t)\frac{dy}{dx} + \underbrace{v(t)\frac{dm(t)}{dt}}_{\text{extra term !}}$$

2-25

Crisis in Circuit Theory

Pre-1970 Definitions of the

3 Basic Circuit Elements

Capacitors, Resistors, and Inductors

give wrong circuit solutions

when the elements are

time-varying or nonlinear

2-26

To Resolve the

Crisis

Capacitors, Resistors, Inductors

must be

redefined

via an

AXIOMATIC APPROACH

2-27

All Results Derived
from An
Axiomatic Approach 2-28
are
Timeless !

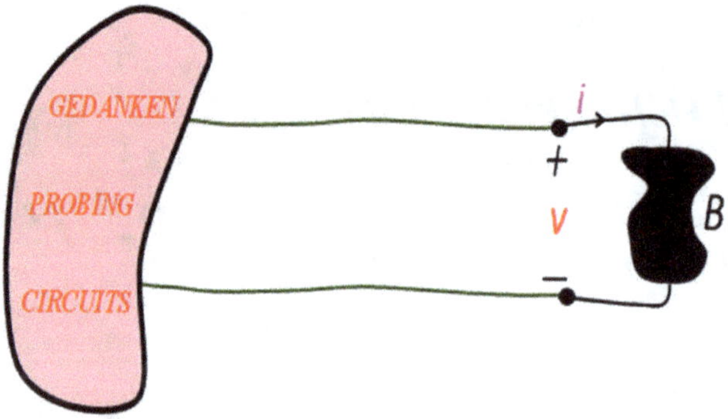

2-29

Axiomatic Definition of
Voltage Across a Pair of Terminals.

The voltage v(t) across a pair of electrical terminals at time t is defined as the number read from a

voltmeter 2-30

connected across the terminals at time t.

The ideal voltmeter has an
infinite resistance.
Attaching a voltmeter across any pair of nodes in an electrical circuit does not disturb the circuit.

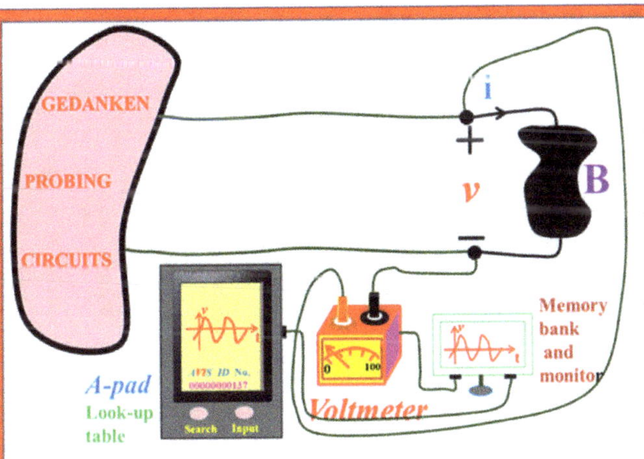

Axiomatic definition of a 2-terminal circuit element

2-31

Axiomatic Definition of Current Flowing Through a Wire in a Circuit.

The current i(t) flowing through a wire at time t is defined as the number read from an

Ammeter

inserted through a wire at time t.

An ammeter has

zero resistance.

Inserting an ammeter in any wire of a circuit does not disturbed the circuit.

2-32

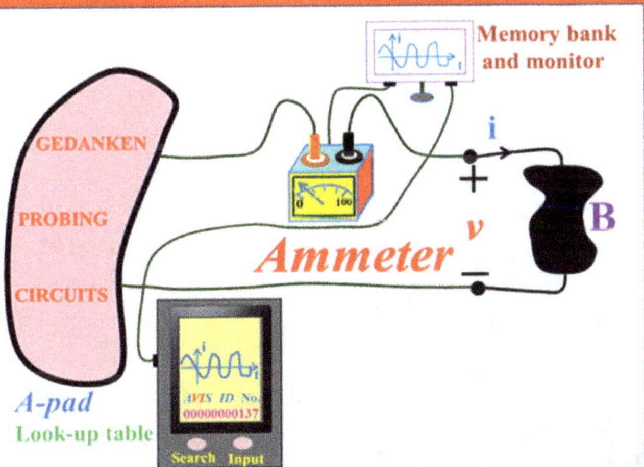

Axiomatic definition of a 2-terminal circuit element

2-33

4 Basic Circuit Variables

- *Voltage* $v(t)$
- *Current* $i(t)$ 2-34
- *Charge* $q(t) \triangleq \int_{-\infty}^{t} i(\tau)d\tau$, or $i = \dfrac{dq(t)}{dt}$
- *Flux* $\varphi(t) \triangleq \int_{-\infty}^{t} v(\tau)d\tau$, or $v = \dfrac{d\varphi(t)}{dt}$

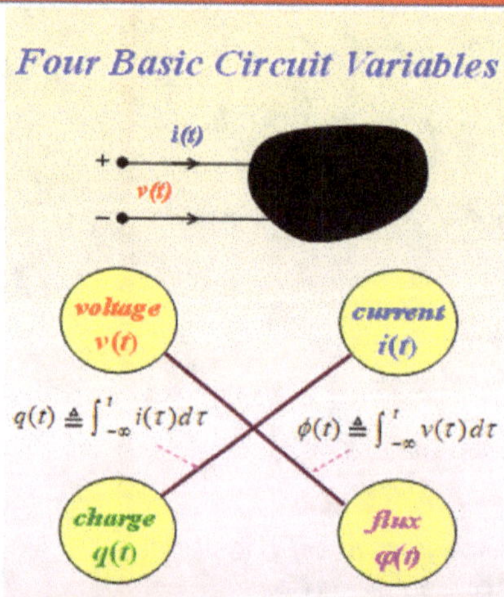

2-35

A 2-terminal Electrical Black Box

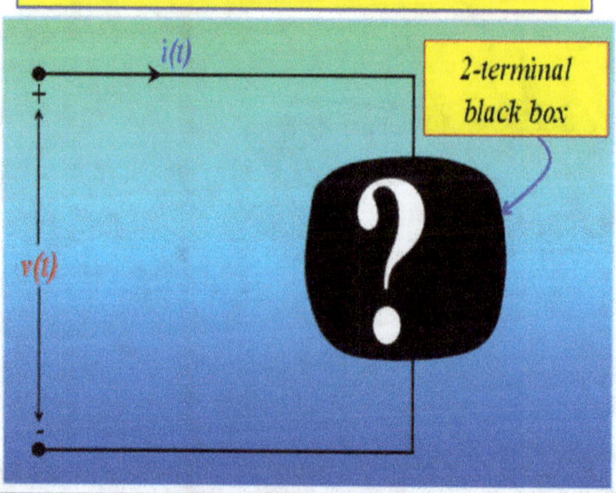

2-36

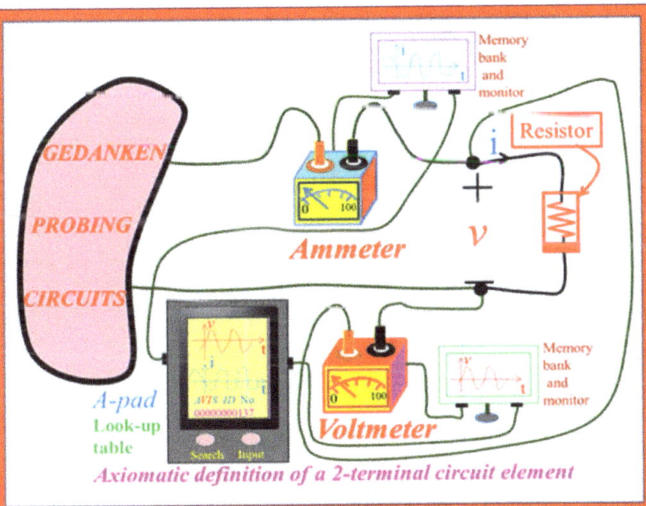

2-37

Constitutive Relation

The characteristic curve $f(x, y)=0$ on which 2 complementary variables x and y must follow at all times of a 2-terminal black box $\mathcal{B}$ called the constitutive relation of $\mathcal{B}$.

2-38

Perfect Device Model

The *perfect device model* is simply an *analog* " *look-up-table* "

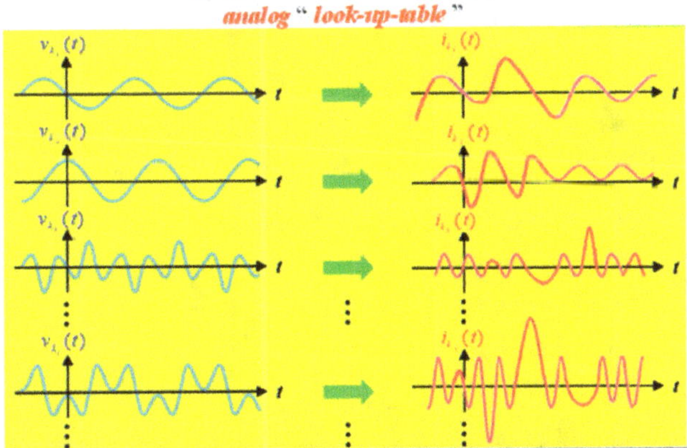

2-39

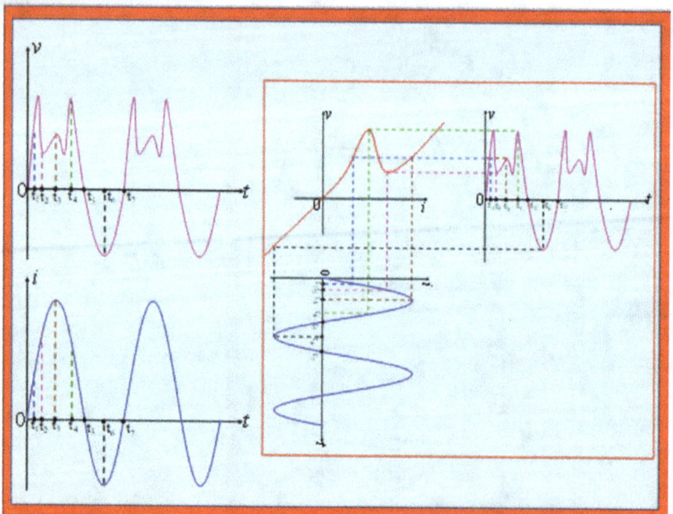

2-40

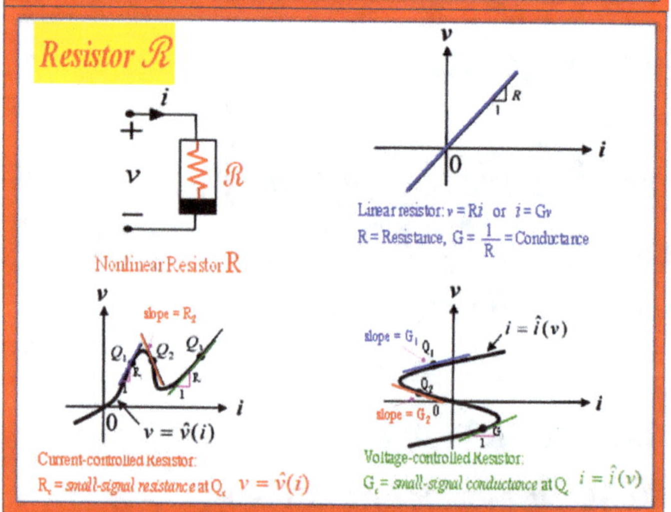
Resistor ℛ
Nonlinear Resistor R
Linear resistor: v = Ri or i = Gv
R = Resistance, G = 1/R = Conductance
slope = R₂
v = v̂(i)
i = î(v)
slope = G₁
slope = G₂
Current-controlled Resistor:
Rₜ = small-signal resistance at Qₜ v = v̂(i)
Voltage-controlled Resistor:
Gₜ = small-signal conductance at Qₜ i = î(v)

2-41

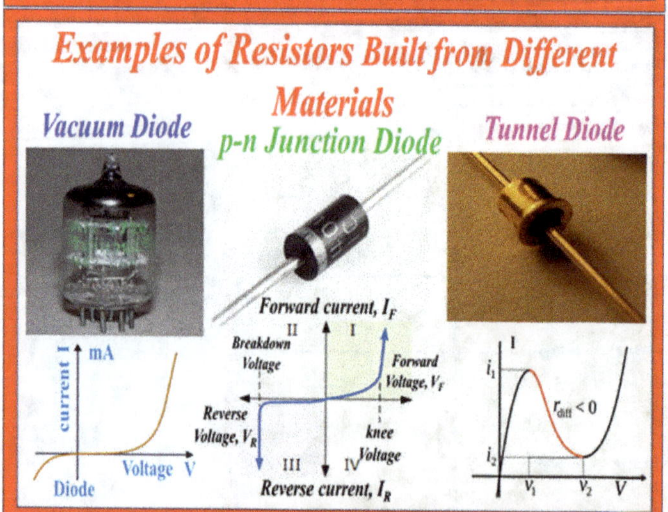
Examples of Resistors Built from Different Materials
Vacuum Diode
p-n Junction Diode
Tunnel Diode
current I
mA
Voltage V
Diode
Forward current, I_F
Breakdown Voltage
Forward Voltage, V_F
Reverse Voltage, V_R
knee Voltage
Reverse current, I_R
r_diff < 0

2-42

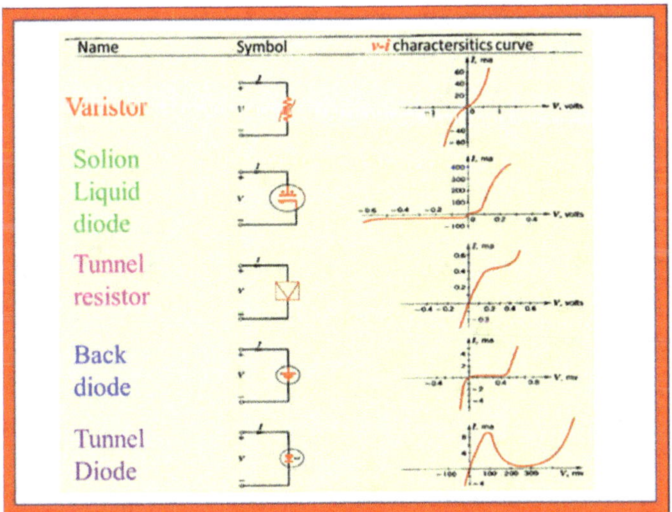

Name
Symbol
v-i characteristics curve
Varistor
Solion Liquid diode
Tunnel resistor
Back diode
Tunnel Diode
I, ma
V, volts

2-43

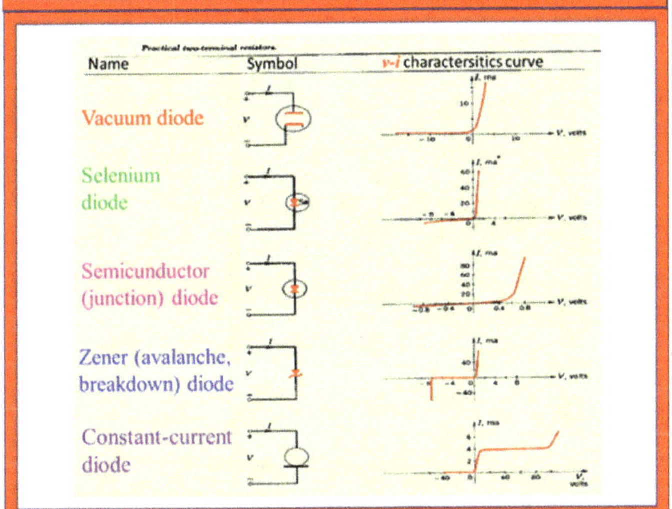

Practical two-terminal resistors.
Name
Symbol
v-i characteristics curve
Vacuum diode
Selenium diode
Semicunductor (junction) diode
Zener (avalanche, breakdown) diode
Constant-current diode
I, ma
V, volts

2-44

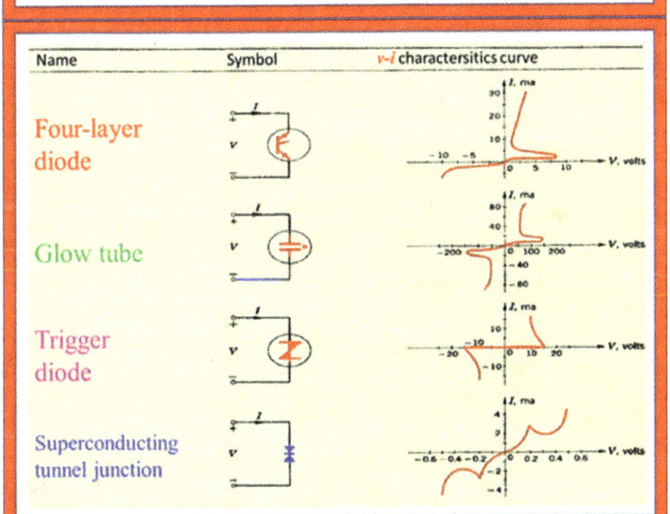

Name
Symbol
v-i characteristics curve
Four-layer diode
Glow tube
Trigger diode
Superconducting tunnel junction
I, ma
V, volts

2-45

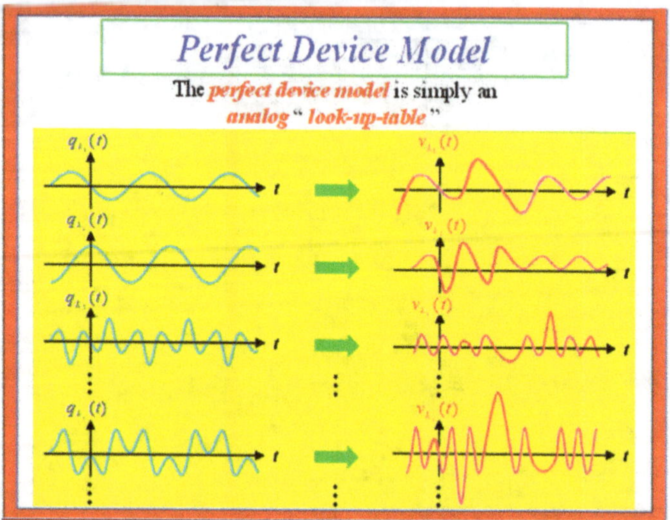
Perfect Device Model
The perfect device model is simply an analog " look-up-table "

2-46

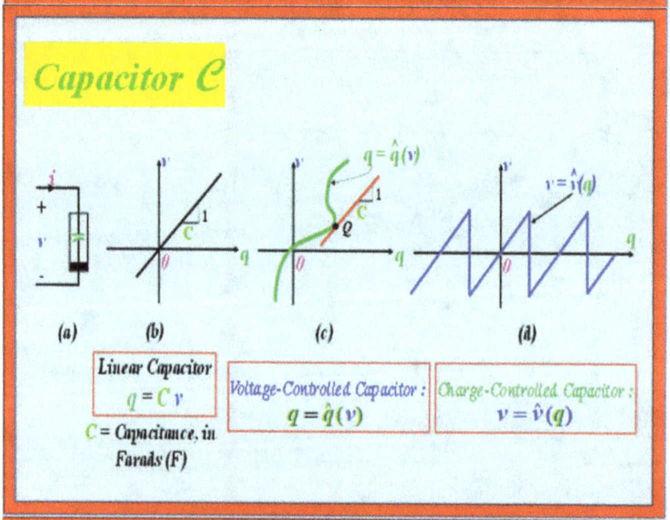
Capacitor C
(a)
(b)
(c)
(d)
Linear Capacitor
q = C v
C = Capacitance, in Farads (F)
Voltage-Controlled Capacitor :
q = q̂(v)
Charge-Controlled Capacitor :
v = v̂(q)

2-47

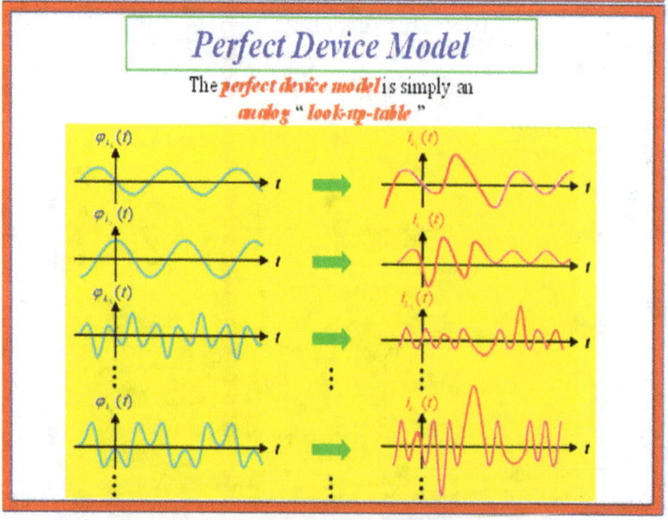
Perfect Device Model
The perfect device model is simply an analog " look-up-table "

2-48

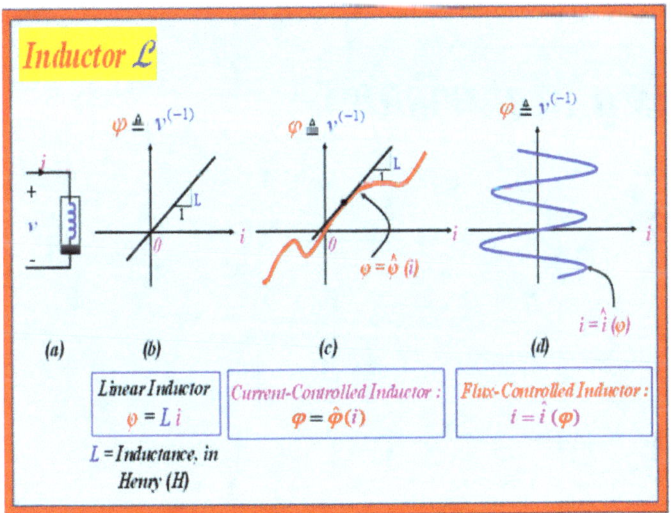

2-49

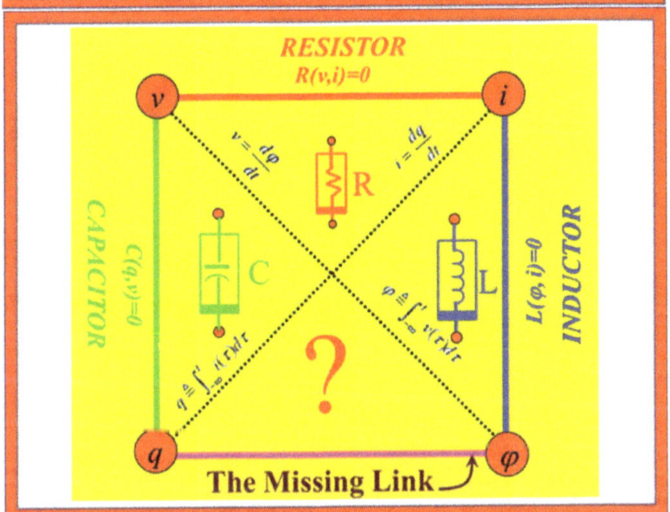

2-50

The missing circuit element is the memristor !

2-51

What is a Memristor ?

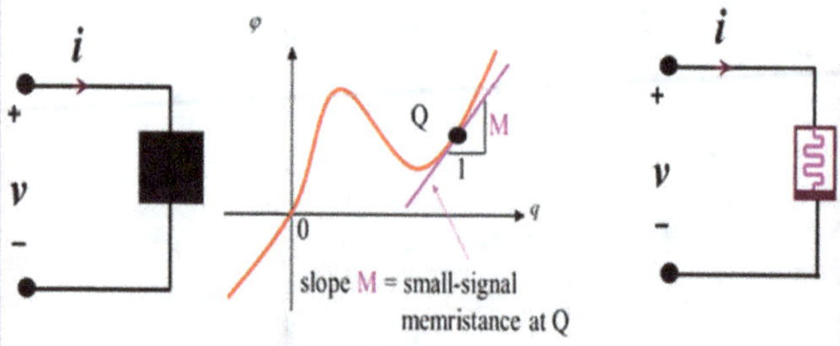

2-52

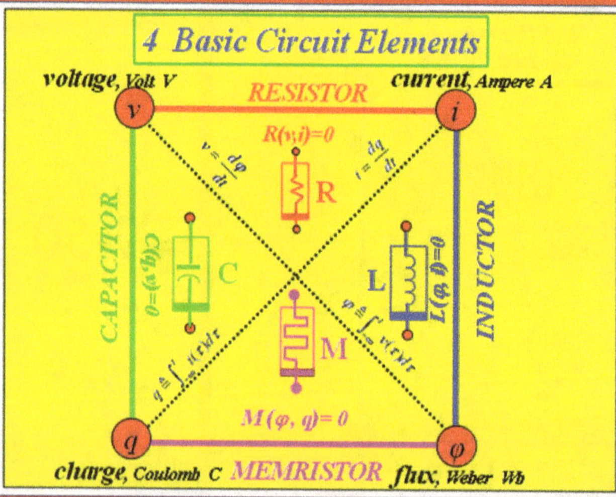

2-53

Memristor

is an acronym for

Memory Resistor

2-54

IEEE TRANSACTIONS ON CIRCUIT THEORY, VOL. CT-18, NO. 5, SEPTEMBER 1971 507

Memristor—The Missing Circuit Element

LEON O. CHUA, SENIOR MEMBER, IEEE

From:
Leon O. Chua,
"Memristor - The Missing Circuit Element,"
IEEE Transaction on Circuit Theory,
Vol. CT-18, No. 5, pp. 507-519, 1971

2-55

1 May 2008

2-56

Nature pages 80–83 May 1, 2008

LETTERS

The missing memristor found

Dmitri B. Strukov[1], Gregory S. Snider[1], Duncan R. Stewart[1] & R. Stanley Williams[1]

2-57

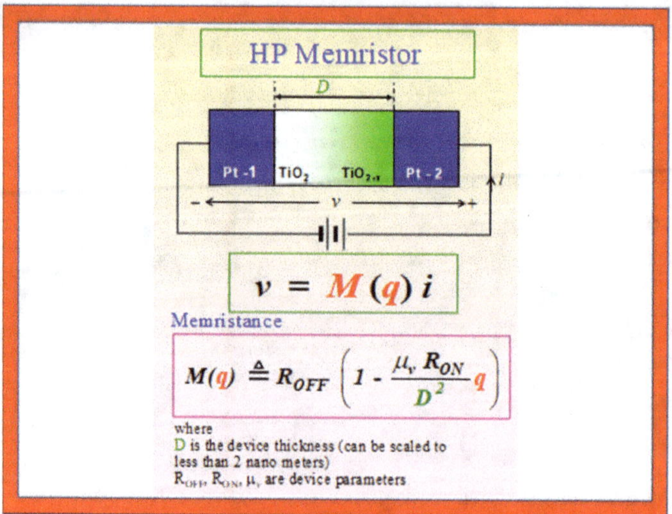
HP Memristor
D
Pt -1
TiO2
TiO2-x
Pt - 2
i
v
v = M (q) i
Memristance
M(q) ≜ R_OFF (1 - μ_v R_ON / D^2 q)
where
D is the device thickness (can be scaled to
less than 2 nano meters)
R_OFF, R_ON, μ_v are device parameters

2-58

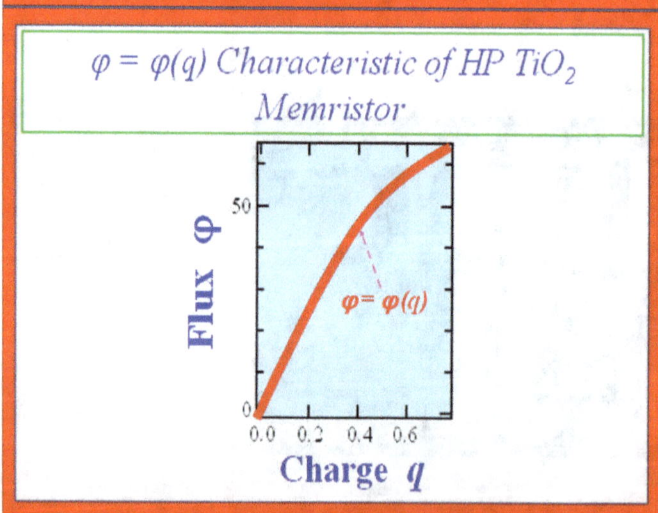
φ = φ(q) Characteristic of HP TiO2 Memristor
Flux φ
50
0
φ= φ(q)
0.0 0.2 0.4 0.6
Charge q

2-59

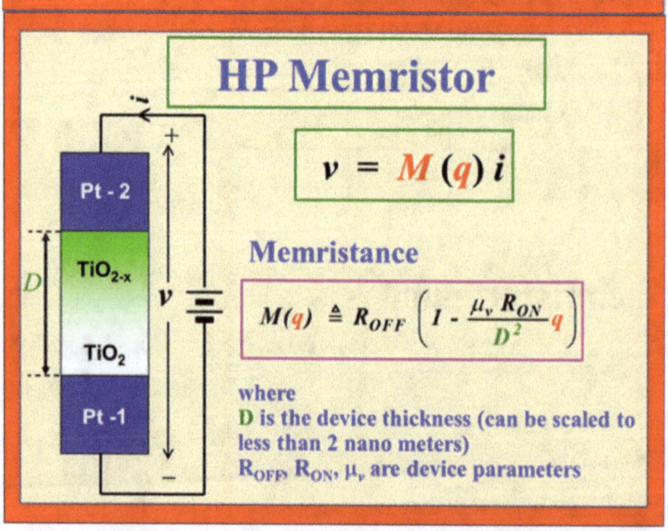
HP Memristor
i
Pt - 2
TiO2-x
TiO2
Pt -1
D
v
v = M (q) i
Memristance
M(q) ≜ R_OFF (1 - μ_v R_ON / D^2 q)
where
D is the device thickness (can be scaled to
less than 2 nano meters)
R_OFF, R_ON, μ_v are device parameters

2-60

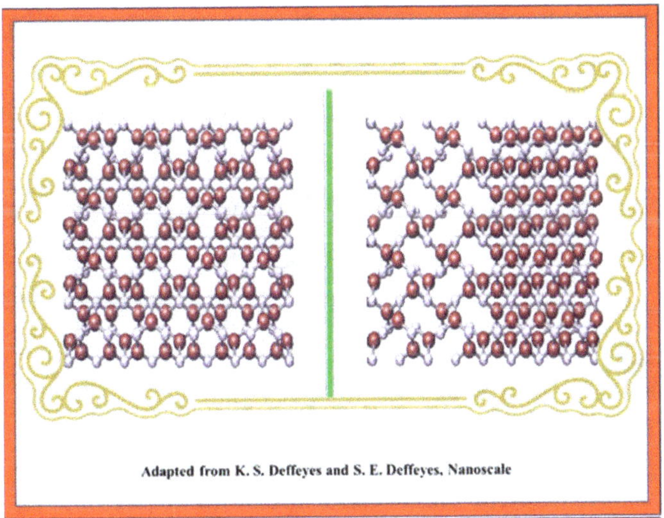
Adapted from K. S. Deffeyes and S. E. Deffeyes, Nanoscale

2-61

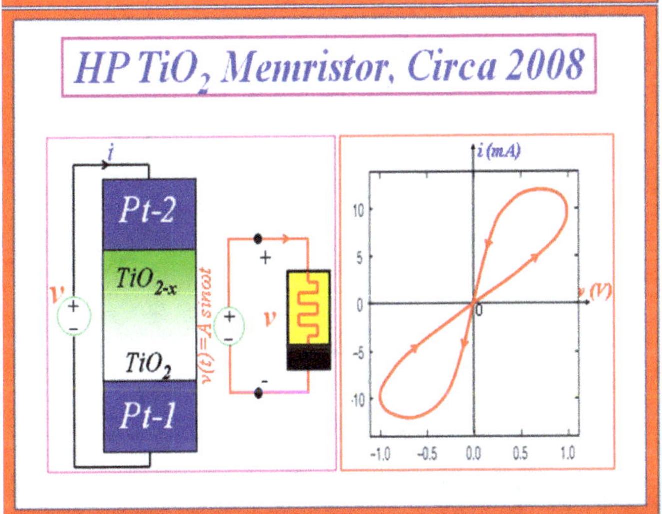
HP TiO2 Memristor, Circa 2008
i
Pt-2
TiO2-x
TiO2
Pt-1
v
v(t)=A sinωt
+
-
v
i (mA)
v (V)
10
5
0
-5
-10
-1.0
-0.5
0.0
0.5
1.0

2-62

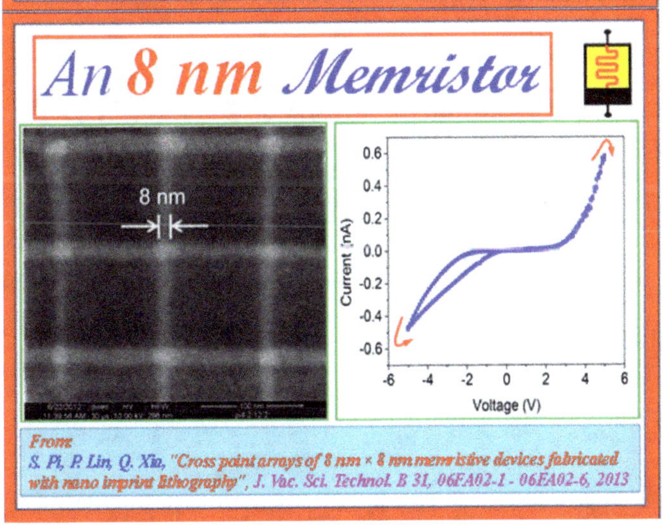
An 8 nm Memristor
8 nm
Current (nA)
Voltage (V)
0.6
0.4
0.2
0.0
-0.2
-0.4
-0.6
-6
-4
-2
0
2
4
6
From:
S. Pi, P. Lin, Q. Xia, "Cross point arrays of 8 nm × 8 nm memristive devices fabricated with nano imprint lithography", J. Vac. Sci. Technol. B 31, 06FA02-1 - 06FA02-6, 2013

2-63

Memristor

made from

A single Layer of the Molecule MoS_2 2-64

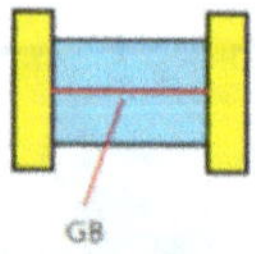

From:
V. K. Sangwan, D. Jariwala, I. S. Kim, K. S. Chen, T. J. Marks, L. J. Lauhon, M. C. Hersam, "Gate-tunable memristive phenomena mediated by grain boundaries in single-layer MoS_2", Nature Nanotechnology 10, p. 403-406, 2015

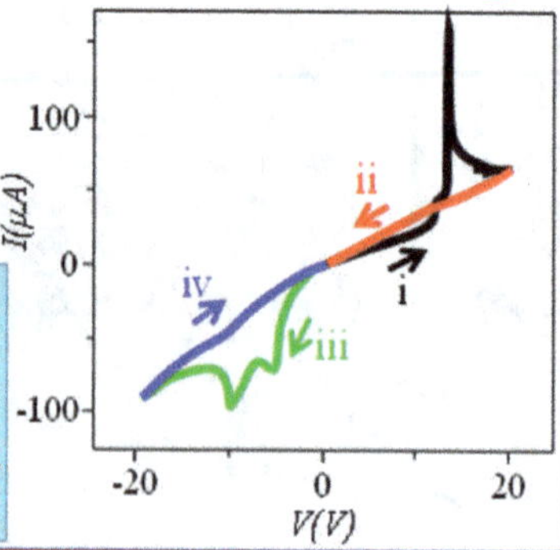

One-Molecule Memristor 2-65

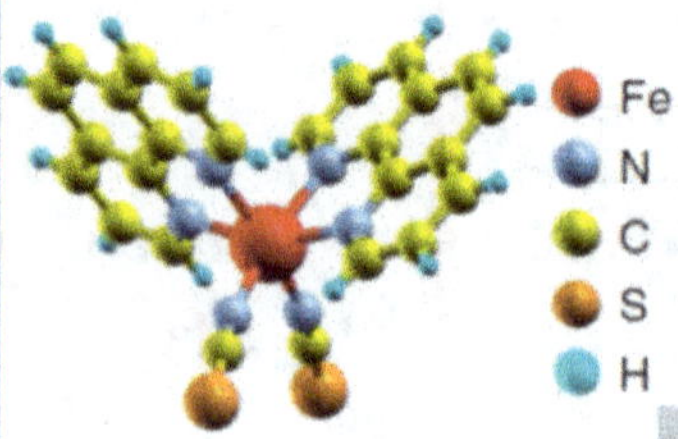

Toshio Miyamachi et al.
" Robust spin crossover and memristance across a single molecule "
Nature Communications 3, 938, p.1-6, 2012

Nature: Molecule Changes Magnetism and Conductance

KIT Press Release No.116, July 2012
Karlsruhe Institute of Technology.

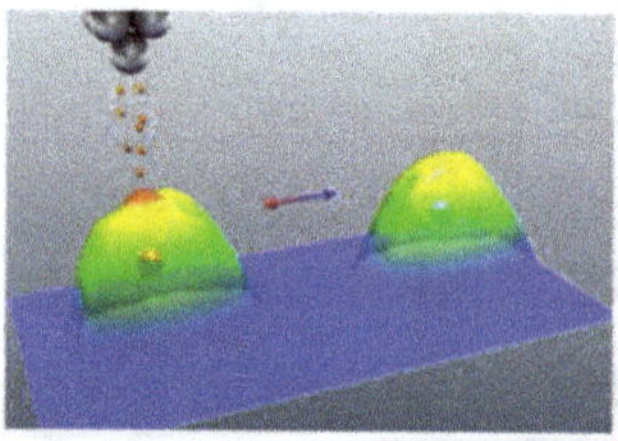

As nano devices shrink below 3 nm,

transistors begins to fail, while the

memristor phenomenon 2-66

begins to dominate, and the circuit dynamics

becomes increasingly nonlinear.

Nature pages 80–83 May 1, 2008

LETTERS

The missing memristor found

Dmitri B. Strukov[1], Gregory S. Snider[1], Duncan R. Stewart[1] & R. Stanley Williams[1]

2-67

Fundamental Theorem on Ideal Memristors

It is
impossible
to make an
Ideal Memristor
using only the 3 basic circuit elements

Capacitor | Resistor | Inductor

2-68

Mendeleev

2-69

Mendeleev's First Published Periodic Table, 1869

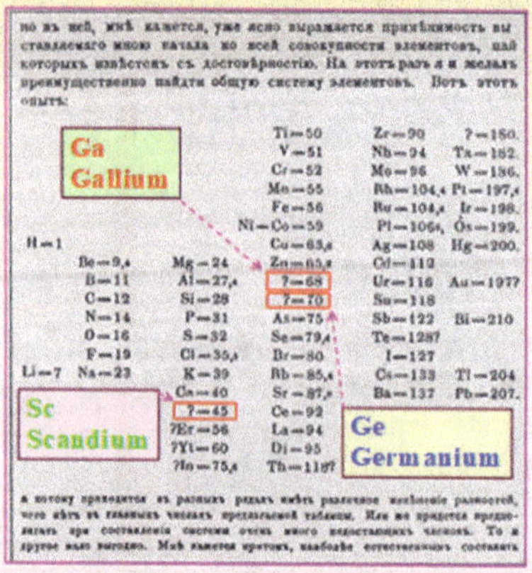

но въ ней, мнѣ кажется, уже ясно выражается примѣнимость выставляемаго мною начала ко всей совокупности элементовъ, пай которыхъ извѣстенъ съ достовѣрностію. На этотъ разъ я и желалъ преимущественно найдти общую систему элементовъ. Вотъ этотъ опытъ:

			Ti—50	Zr—90	?—180.
			V—51	Nb—94	Ta—182.
			Cr—52	Mo—96	W—186.
			Mn—55	Rh—104,4	Pt—197,4
			Fe—56	Ru—104,4	Ir—198.
			Ni—Co—59	Pl—106,6	Os—199.
H—1			Cu—63,4	Ag—108	Hg—200.
	Be—9,4	Mg—24	Zn—65,2	Cd—112	
	B—11	Al—27,4	?—68	Ur—116	Au—197?
	C—12	Si—28	?—70	Sn—118	
	N—14	P—31	As—75	Sb—122	Bi—210
	O—16	S—32	Se—79,4	Te—128?	
	F—19	Cl—35,5	Br—80	I—127	
Li—7	Na—23	K—39	Rb—85,4	Cs—133	Tl—204
		Ca—40	Sr—87,6	Ba—137	Pb—207.
		?—45	Ce—92		
		?Er—56	La—94		
		?Yt—60	Di—95		
		?In—75,6	Th—118?		

а потому приходится въ разныхъ рядахъ имѣть различное измѣненіе разностей, чего нѣтъ въ главныхъ числахъ предлагаемой таблицы. Или же придется предполагать при составленіи системы очень много недостающихъ членовъ. То и другое мало выгодно. Мнѣ кажется притомъ, наиболѣе естественнымъ составить

2-70

Our

4 Circuit-Element

Axiomatic

Approach

dates back to

Aristotle !

Aristotle, 350 BC

2-71

What are the Building Blocks of Matter ?

2-72

2-73

Aristotle's Theory of Matter

All ***matter*** consisted of ***four elements***:

1. EARTH 2. WATER 3. AIR 4. FIRE

Each of these elements exhibited two of ***four fundamental properties*** :

1. Moistness 2. Dryness 3. Coldness 4. Hotness

2-74

Aristotle's 4 Pairs of Complementary Variables

- (Dryness, Hotness) ⟹ FIRE
- (Moistness, Hotness) ⟹ AIR
- (Moistness, Coldness) ⟹ WATER
- (Dryness, Coldness) ⟹ EARTH

2-75

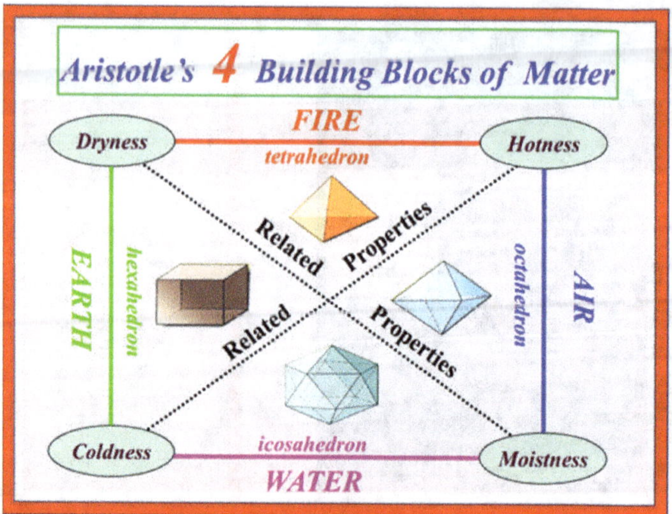
Aristotle's 4 Building Blocks of Matter
FIRE
tetrahedron
Dryness
Hotness
EARTH
hexahedron
AIR
octahedron
Related Properties
Related Properties
Coldness
Moistness
icosahedron
WATER

2-76

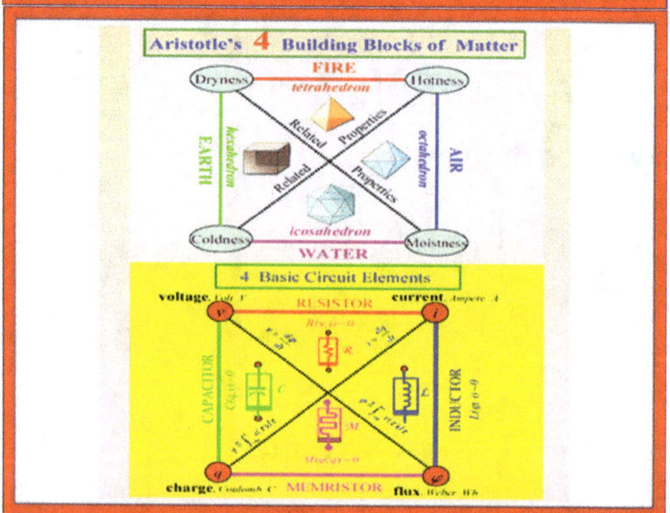
Aristotle's 4 Building Blocks of Matter
FIRE
tetrahedron
Dryness
Hotness
EARTH
hexahedron
AIR
octahedron
Related Properties
Related Properties
Coldness
Moistness
icosahedron
WATER
4 Basic Circuit Elements
voltage
RESISTOR
current
CAPACITOR
INDUCTOR
charge
MEMRISTOR
flux

2-77

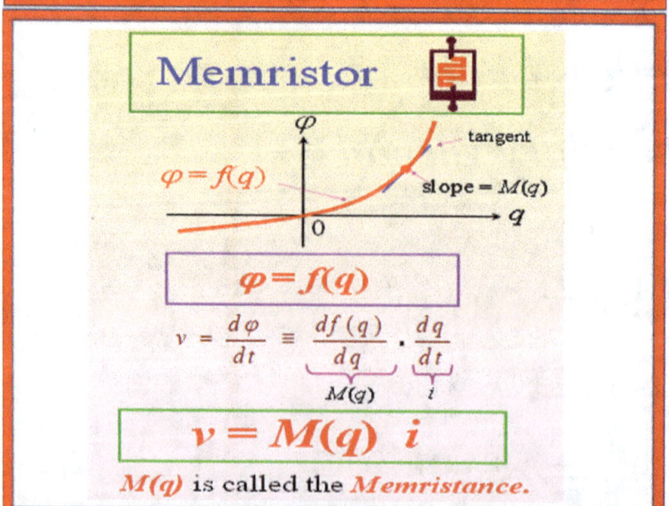
Memristor
φ
tangent
φ = f(q)
slope = M(q)
0
q
φ = f(q)
v = dφ/dt = df(q)/dq · dq/dt
M(q)
i
v = M(q) i
M(q) is called the Memristance.

2-78

Memristor

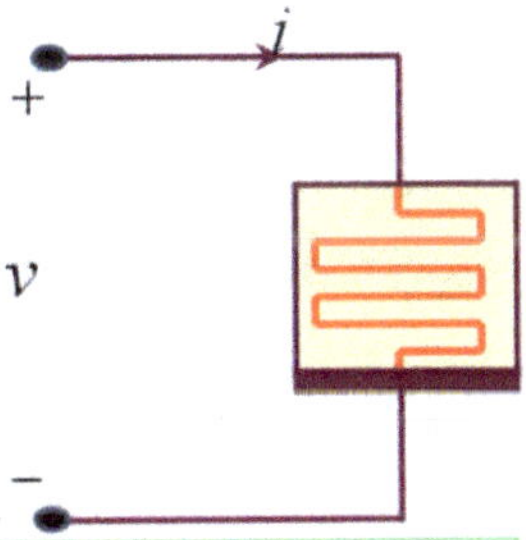

$$v = M(q)\ i$$

$$\dot{q} = i$$

2-79

Memristor

is defined by a

State - Dependent

Ohm's Law

2-80

How To Recover Memristor Constitutive Relation $\varphi = \hat{\varphi}(q)$ From Its Memristance $M(q)$?

$$v = M(q)\, i$$

$$\dot{q} = i$$

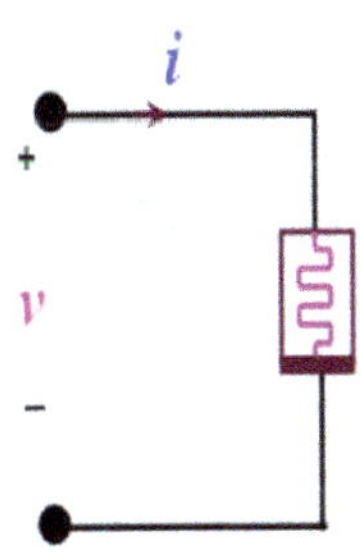

Constitutive Relation :

$$\varphi = \int M(q)\, dq \triangleq \hat{\varphi}(q)$$

where $\hat{\varphi}(q)$ is unique to within an arbitrary constant.

2-81

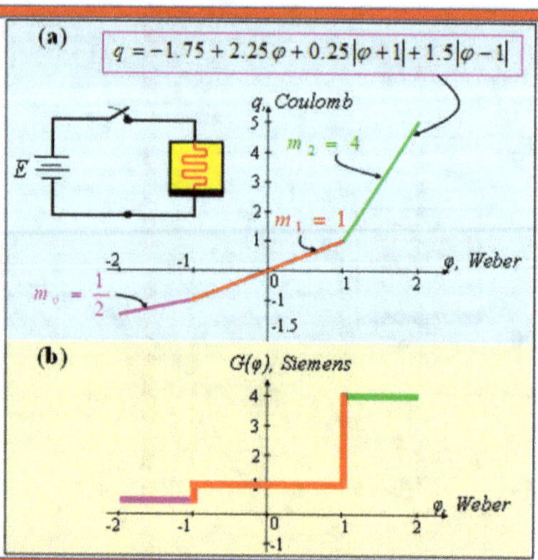

2-82

Universal Formula for Continuous PWL Functions

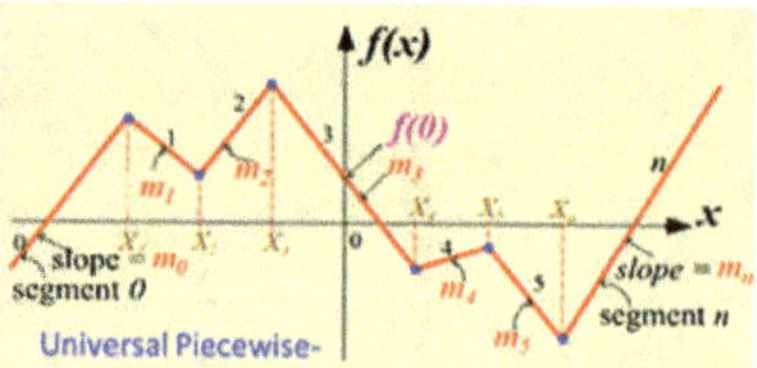

Universal Piecewise-Linear Formula

Every continuous Piecewise-Linear Function can be described by the following exact formula

$$y = f(x) = a_0 + a_1 x + \sum_{j=1}^{n} b_j |x - X_j|$$

where

$$a_1 = \frac{1}{2}(m_0 + m_n)$$

$$b_j = \frac{1}{2}(m_j - m_{j-1})$$

$$a_0 = f(0) - \sum_{j=1}^{n} b_j |X_j|$$

and $X_1, X_2, \ldots\ldots X_n$ are the coordinates of X_j at the n-1 breakpoints

Half Sum Half difference, Then Null Formula

2-83

Mechanical Analogs of Memristors

2-84

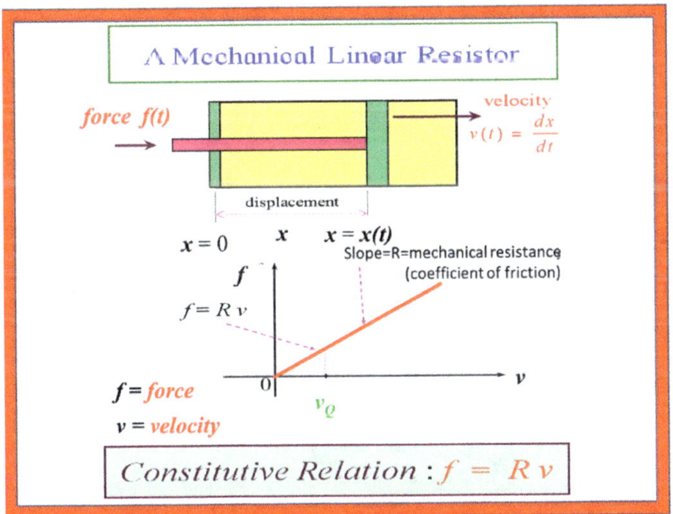
A Mechanical Linear Resistor
force f(t)
velocity
v(t) = dx/dt
displacement
x = 0
x
x = x(t)
Slope=R=mechanical resistance
(coefficient of friction)
f
f = R v
0
v
f = force
v = velocity
v_Q
Constitutive Relation : f = R v

2-85

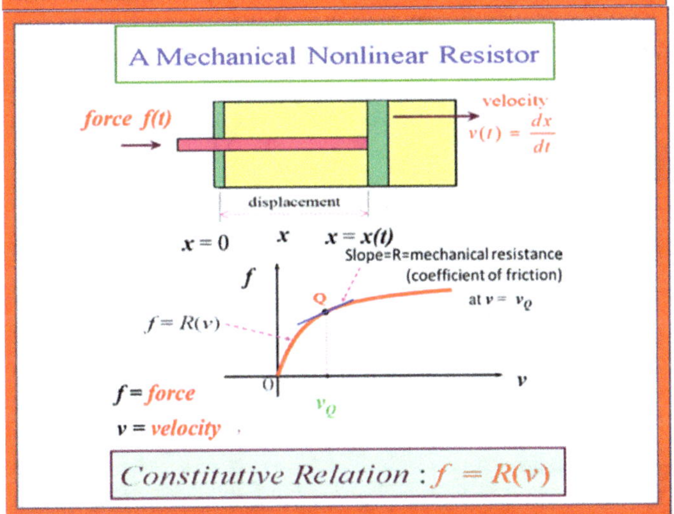
A Mechanical Nonlinear Resistor
force f(t)
velocity
v(t) = dx/dt
displacement
x = 0
x
x = x(t)
Slope=R=mechanical resistance
(coefficient of friction)
at v = v_Q
f
Q
f = R(v)
0
v
f = force
v = velocity
v_Q
Constitutive Relation : f = R(v)

2-86

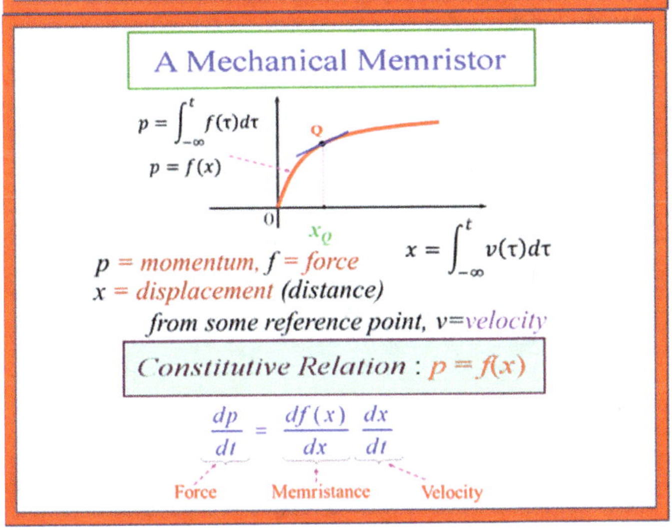
A Mechanical Memristor
$p = \int_{-\infty}^{t} f(\tau)d\tau$
Q
$p = f(x)$
0
x_Q
p = momentum, f = force
$x = \int_{-\infty}^{t} v(\tau)d\tau$
x = displacement (distance)
from some reference point, v=velocity
Constitutive Relation : p = f(x)
$\frac{dp}{dt} = \frac{df(x)}{dx}\frac{dx}{dt}$
Force
Memristance
Velocity

2-87

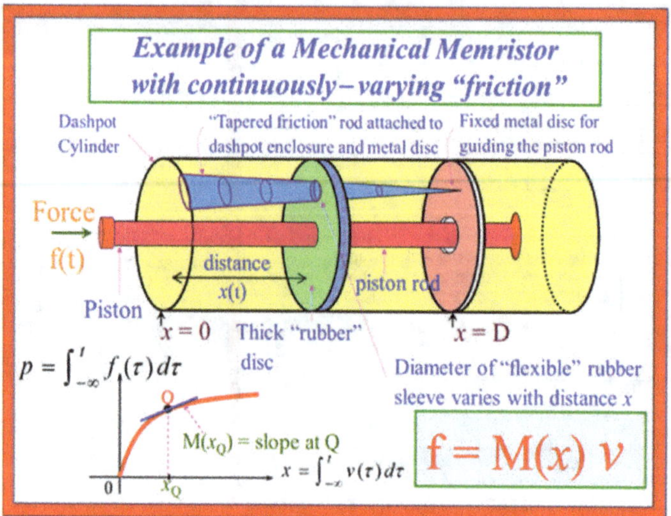
Example of a Mechanical Memristor with continuously-varying "friction"
Dashpot Cylinder
"Tapered friction" rod attached to dashpot enclosure and metal disc
Fixed metal disc for guiding the piston rod
Force
f(t)
distance
x(t)
piston rod
Piston
x = 0
Thick "rubber" disc
x = D
Diameter of "flexible" rubber sleeve varies with distance x
$p = \int_{-\infty}^{t} f(\tau)\,d\tau$
Q
$M(x_Q)$ = slope at Q
$x = \int_{-\infty}^{t} v(\tau)\,d\tau$
0
x_Q
f = M(x) v

2-88

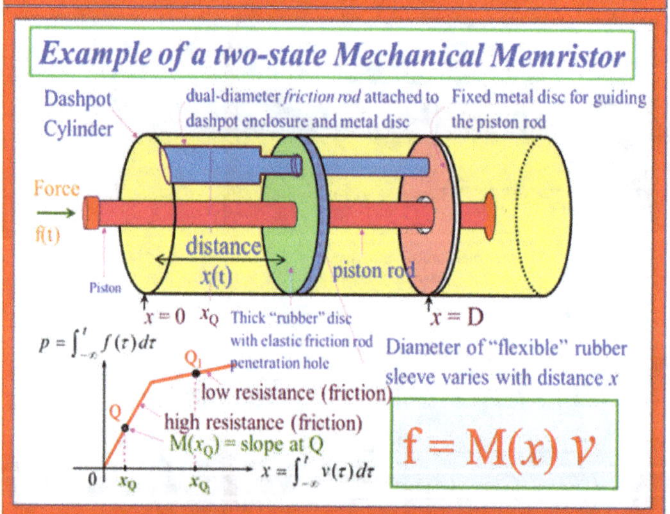
Example of a two-state Mechanical Memristor
Dashpot Cylinder
dual-diameter friction rod attached to dashpot enclosure and metal disc
Fixed metal disc for guiding the piston rod
Force
f(t)
distance
x(t)
piston rod
Piston
x = 0
x_Q
Thick "rubber" disc with elastic friction rod penetration hole
x = D
Diameter of "flexible" rubber sleeve varies with distance x
$p = \int_{-\infty}^{t} f(\tau)\,d\tau$
Q_1
low resistance (friction)
Q
high resistance (friction)
$M(x_Q)$ = slope at Q
$x = \int_{-\infty}^{t} v(\tau)\,d\tau$
0
x_Q
x_{Q_1}
f = M(x) v

2-89

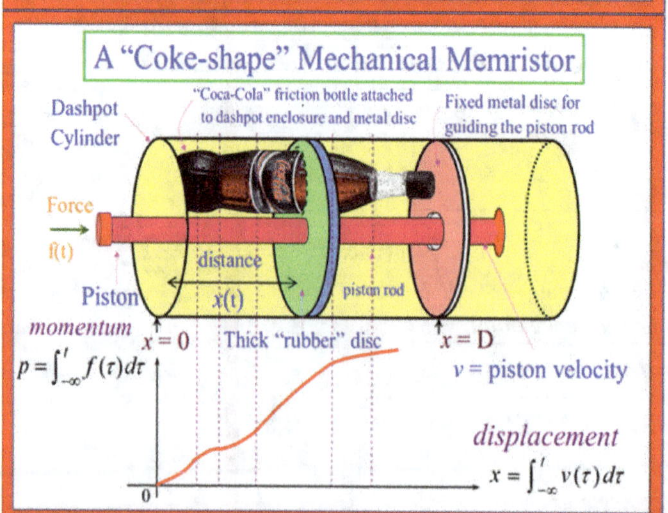
A "Coke-shape" Mechanical Memristor
Dashpot Cylinder
"Coca-Cola" friction bottle attached to dashpot enclosure and metal disc
Fixed metal disc for guiding the piston rod
Force
f(t)
distance
x(t)
piston rod
Piston
momentum
$p = \int_{-\infty}^{t} f(\tau)\,d\tau$
x = 0
Thick "rubber" disc
x = D
v = piston velocity
displacement
$x = \int_{-\infty}^{t} v(\tau)\,d\tau$
0

2-90

What Does Non-Volatile Mean ?

2-91

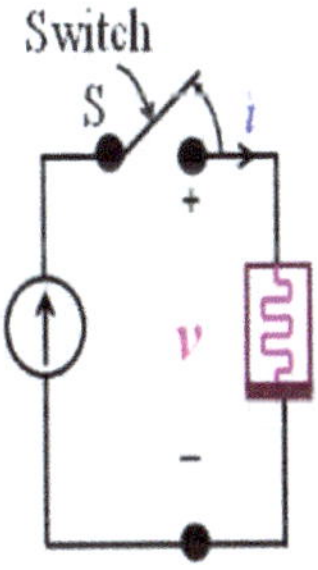

Charge-Controlled Memristor

$$v = M(q)i$$

$$\frac{dq}{dt} = i$$

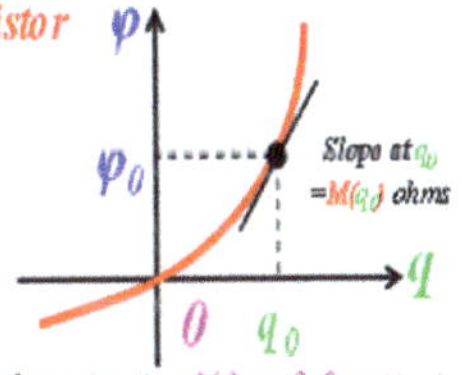

when switch S is open at time $t = t_0$, $i(t) = 0$ *for* $t \geq t_0$.

$$\frac{dq}{dt} = 0$$

$$q(t_0) = \int_{-\infty}^{t_0} i(\tau)d\tau \triangleq q_0$$

where q_0 *is the net accumulated charge from time* $t = -\infty$ *to* $t = t_0$.

$M(t) = M(q_0)$ *for* $t \geq t_0$

Memristive Logic is Coded by Resistance

2-92

As with *Shannon's original formulation* of digital logic, the physical representation for the logic value of a

Memristor

is either

a high resistance, for binary 0

or

a low resistance for binary 1

Resistance Switching

implies

Memristor

2-93

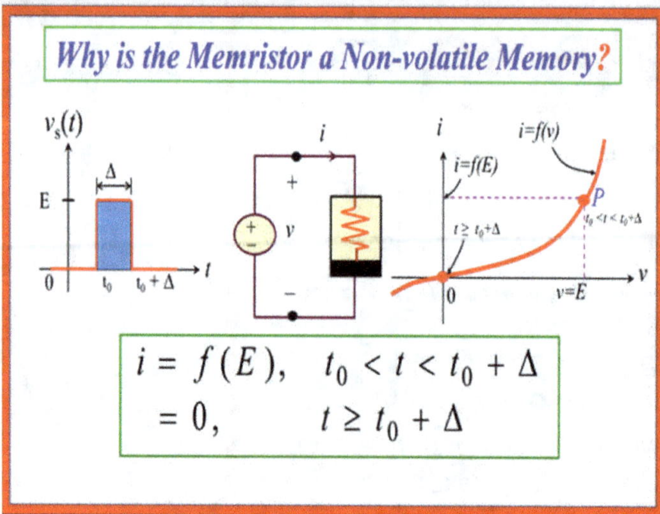
Why is the Memristor a Non-volatile Memory?
$v_s(t)$
E
Δ
0
t_0
$t_0+\Delta$
t
i
+
v
−
i
i=f(v)
i=f(E)
P
$t_0<t<t_0+\Delta$
$t\geq t_0+\Delta$
0
v=E
v
$i = f(E), \quad t_0 < t < t_0 + \Delta$
$= 0, \quad t \geq t_0 + \Delta$

2-94

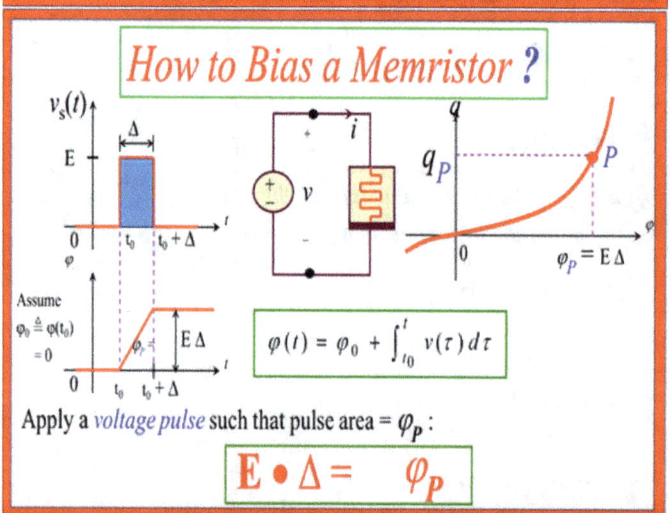
How to Bias a Memristor ?
$v_s(t)$
E
Δ
0
t_0
$t_0+\Delta$
t
φ
Assume
$\varphi_0 \triangleq \varphi(t_0) = 0$
φ_P
E Δ
+
i
v
−
q
q_P
P
0
$\varphi_P = E\Delta$
φ
$\varphi(t) = \varphi_0 + \int_{t_0}^{t} v(\tau)\,d\tau$
Apply a voltage pulse such that pulse area = φ_P :
$E \bullet \Delta = \varphi_P$

2-95

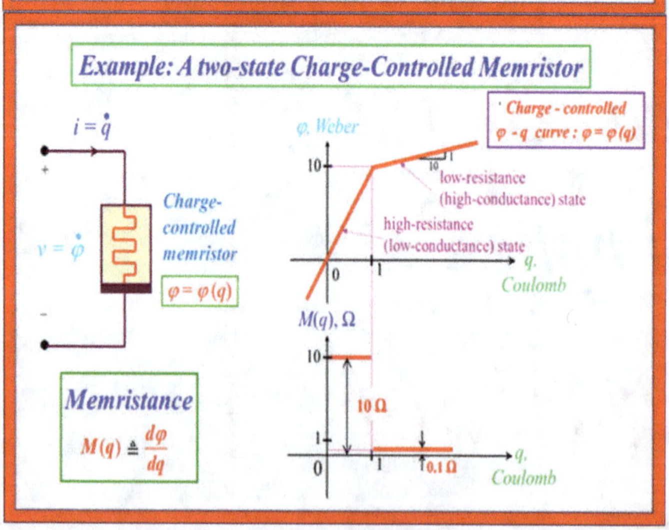
Example: A two-state Charge-Controlled Memristor
$i=\dot{q}$
+
$v=\dot{\varphi}$
−
Charge-controlled memristor
$\varphi=\varphi(q)$
Memristance
$M(q) \triangleq \frac{d\varphi}{dq}$
φ, Weber
Charge - controlled φ - q curve : φ = φ(q)
10
low-resistance (high-conductance) state
high-resistance (low-conductance) state
0
1
q, Coulomb
M(q), Ω
10
10 Ω
1
0
1
0.1 Ω
q, Coulomb

2-96

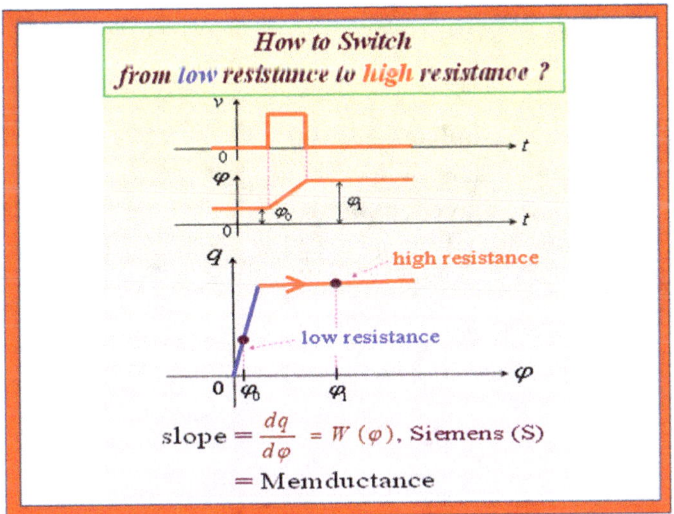
How to Switch
from low resistance to high resistance ?
high resistance
low resistance
slope = $\frac{dq}{d\varphi}$ = $W(\varphi)$, Siemens (S)
= Memductance

2-97

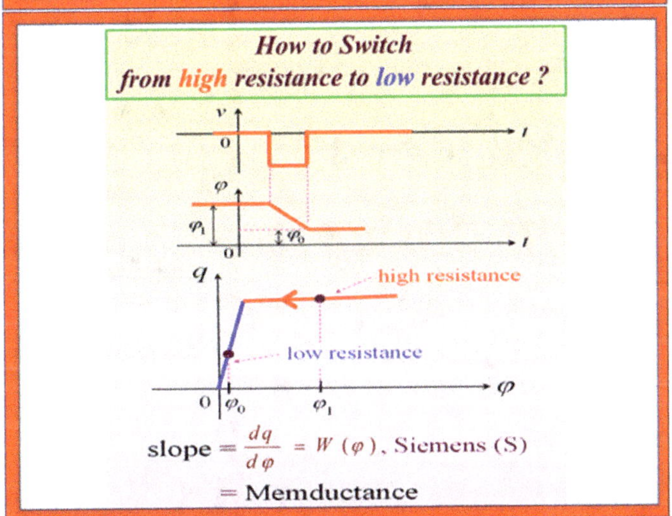
How to Switch
from high resistance to low resistance ?
high resistance
low resistance
slope = $\frac{dq}{d\varphi}$ = $W(\varphi)$, Siemens (S)
= Memductance

2-98

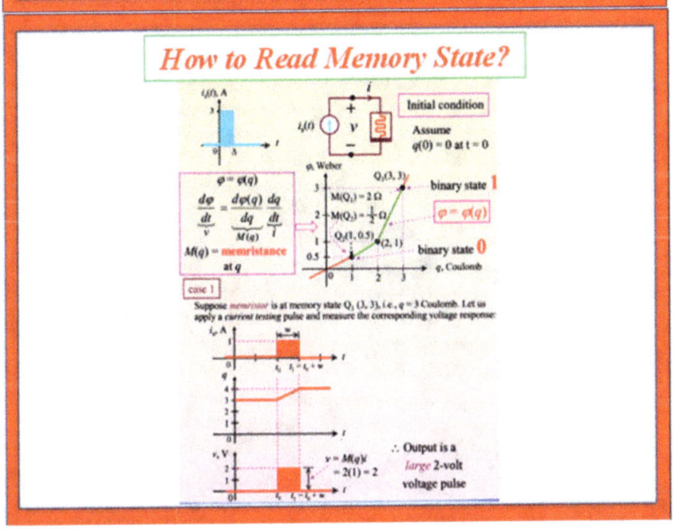
How to Read Memory State?
Initial condition
Assume
binary state 1
binary state 0
case 1
∴ Output is a
large 2-volt
voltage pulse

2-99

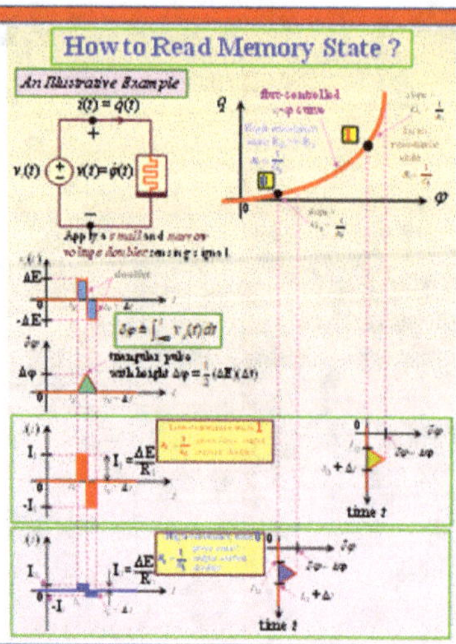

Enlargement of this slide is given in Slides **3-143 and 3-144** 2-100

Non-volatile memories

are estimated to be a

250 billion dollar

Industry by 2020 ! 2-101

Imagine a PC which turns on instantly !

Flash is Short Term Memory

Flash memories are not permanent !

2-102

They lose about 20 per cent of information per decade.

Never

Store your

Family pictures 2-103

on

flash memory sticks

Why not Flash ?

- *Can not* be scaled *below 20 nanometers*
- Poor Retention time: *Fails* after switching between *10,000 and 100,000 times* 2-104
- Low Speed
- Power Hungry

Non-Volatile Nano Memristors

will eventually replace the following conventional computer memories

- ***Flash Memories*** 2-105

- ***DRAMs***

- ***Hard Drives***

B. D. Josephson,
"Possible New Effects in Superconducting Tunneling,"
Physics Letters,
vol. 1, no. 7, pp. 251-253, 1962 2-106

Modeling Quasi-Particle Pair Interference Current in Josephson Junctions

In the quantum-mechanical analysis of the Josephson junction, a small contribution to the device current is derived by Josephson to be given by

$$i = M(\cos \varphi)\, v$$ 2-107

where ***M*** *is a constant which depends on the device parameters. This equation, which usually negligible, represents a* ***memristor*** *defined by*

$$q = M \sin \varphi$$

Brian Josephson
1973 Nobel Prize in Physics:

JOSEPHSON JUNCTION CIRCUIT MODEL

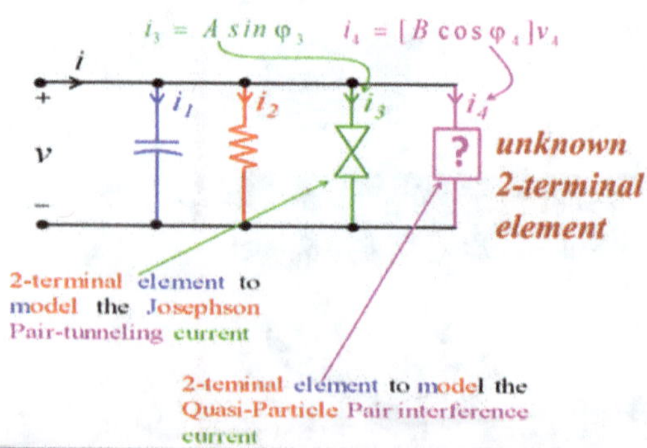

2-108

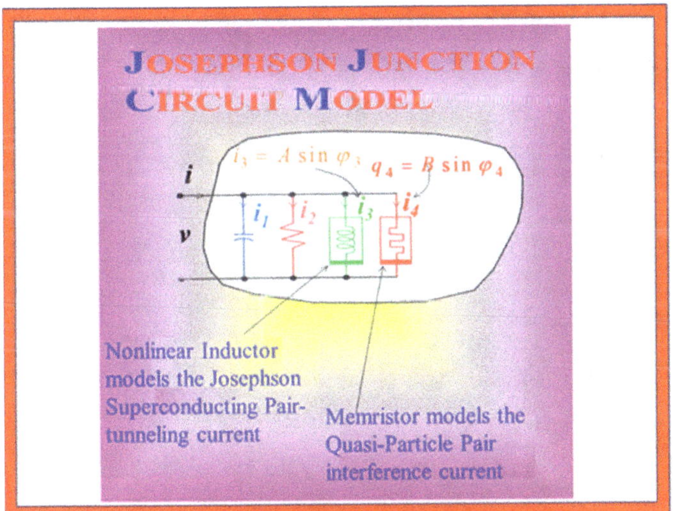

2-109

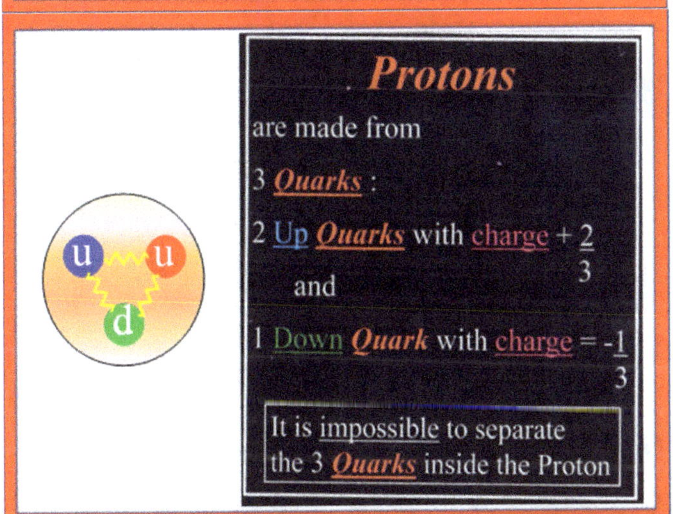

2-110

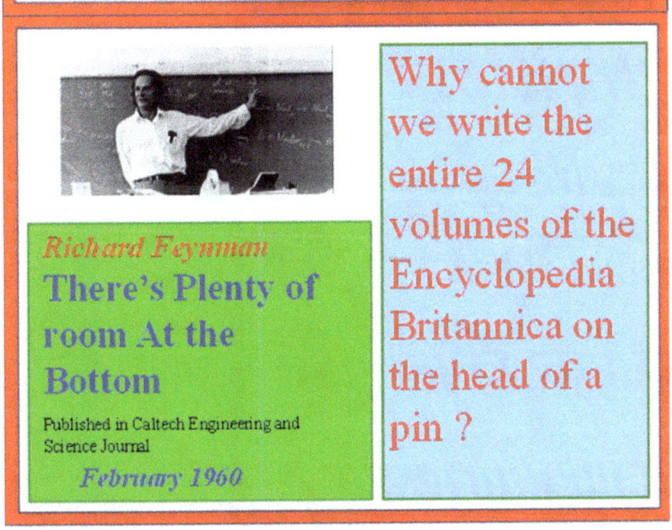

2-111

Richard Feynman

There's Plenty of smarts At the Bottom

Published in Caltech Engineering and Science Journal

February 1960

Why cannot we write the entire 24 volumes of the Encyclopedia Britannica on the head of a pin ? 2-112

Conventional

***Electronic memories** are*

cast in stones

They have no minds ! 2-113

Memristors

are

much more

versatile than

dumb memories ! 2-114

Memristor : New Device with Intelligence

2-115

Memristor

are *not* just *Dumbed Memories*

they are the *RIGHT STUFF* for building

Brainlike Computers

and

Intelligent Machines

2-116

Toy Examples Of Memristor Learning

2-117

Aplysia with a Nobel Prize Medal 2-118

Eric R. Kandel

Nobel Prize in Physiology or Medicine, 2000. 2-119

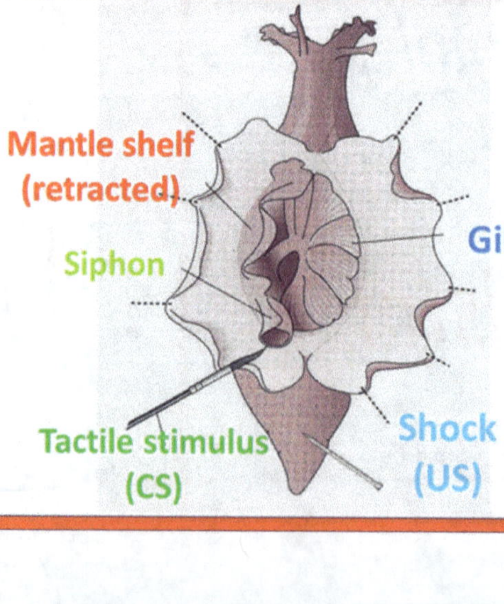

2-120

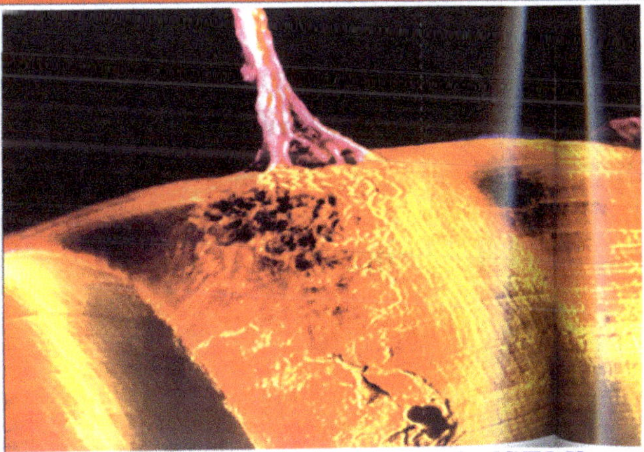

Nerve terminating on muscle [SEM] 2-121

Gill and Siphon Withdrawal Reflex

The gill and its tube-like extension, called the siphon, form a very delicate organ that can be easily damaged. Therefore, when there are signs of danger, the aplysia would retract the gill and siphon inside a protective mantle shelf. 2-122

Example of Déjà vu Response of the Aplysia

2-123

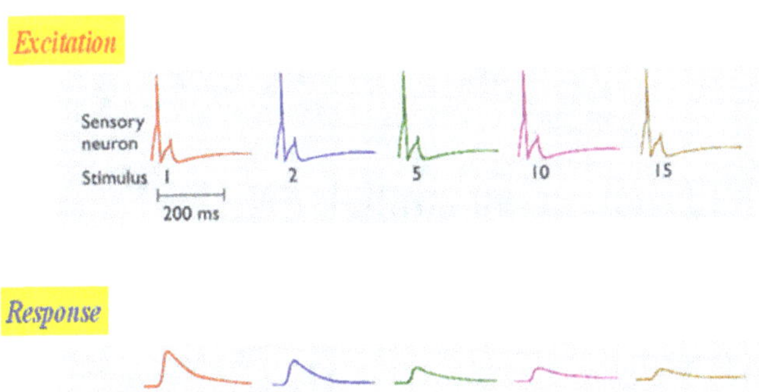

Habituation

Perhaps the simplest form of ***learning*** is ***habituation***. All of us use habituation every day to help us learn *not* to attend or respond to irrelevant stimuli. 2-124

Déjà vu response

Déjà vu response is ***learning*** to ***recognize*** and ***ignore*** benign and boring stimulus. 2-125

Déjà vu Response is Universal

The ***déjà vu response*** (***habituation***) mechanism is the same in all animals with ***nervous*** systems from simple invertebrates to mammals. 2-126

2-127

Habituation

If a drop of water falls on the surface of the sea just over the flower-like disc of a *sea-anemone*, the whole animal contracts vigorously. If, then, a second drop falls within a few minutes of the first, there is less contraction, and finally, on the third and fourth drop, the response disappears altogether. This example clearly exhibited one of the most persuasive phenomena of the animal kingdom, called *habituation*, defined as a *decrease* in the *strength of response with repeated stimulation*. Almost every species from *amoeba* to *man*, exhibits *habituation* when the stimulus is frequently repeated or constantly applied.

2-128

State-Dependent Ohm's Law :

$$i = G(\varphi)\, v$$

State Equation :

$$\frac{d\varphi}{dt} = v$$

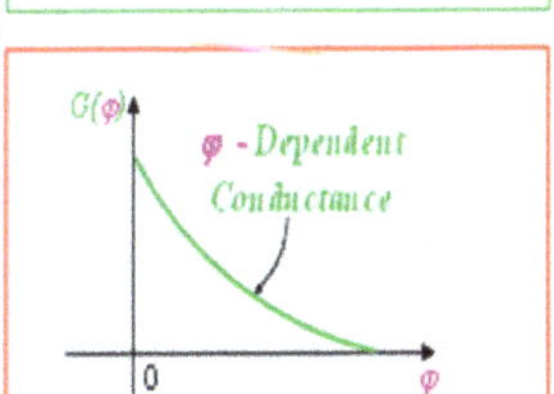

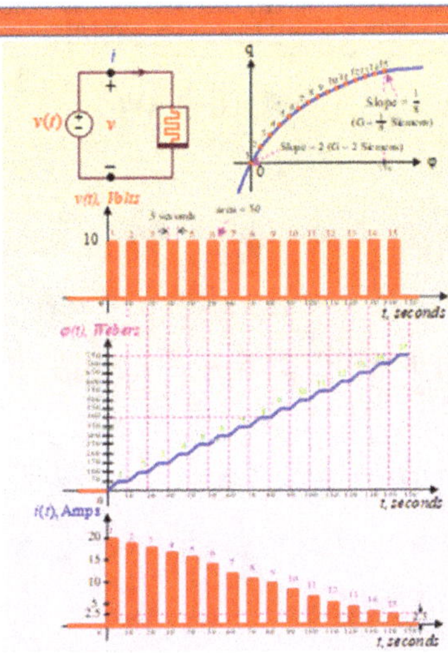

2-129

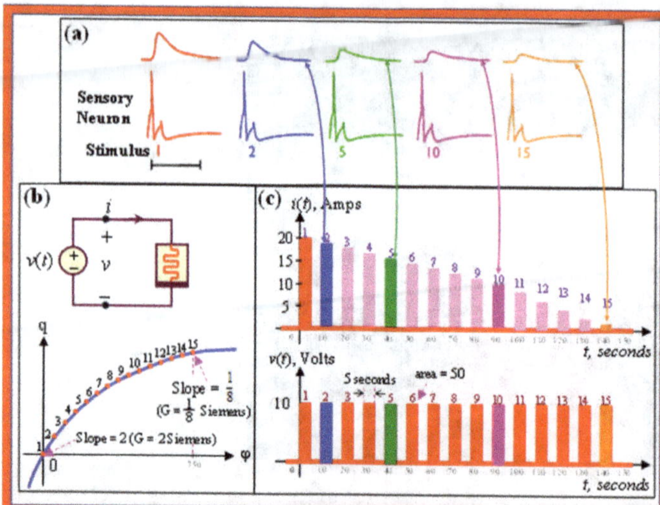

2-130

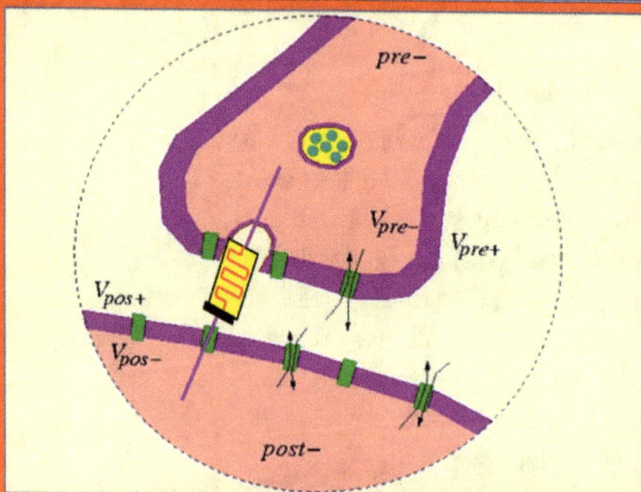

2-131

Memristance is Synaptic Weight

Tuning the ***synaptic weight*** *during* ***learning*** *is equivalent to* ***biasing*** *the operating point with the corresponding* ***slope****.* 2-132

Synapses are Memristors 2-133

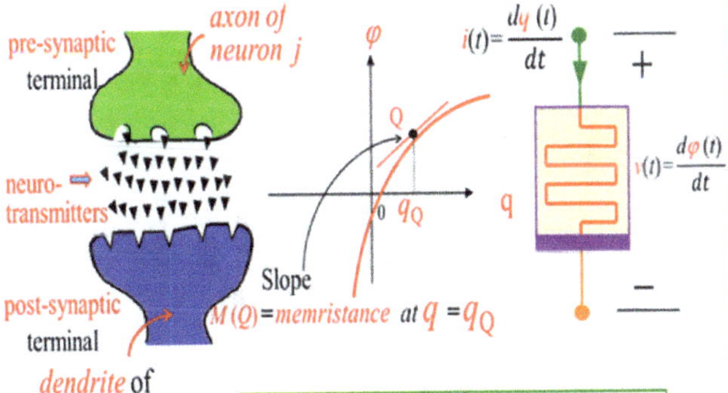

Synaptic strength ≡ memristance

Synapses are *Memristors!* 2-134

LTP

In 1973, Bliss and Lomo found that a *tetanus* (a short high-frequency pulse train) 2-135

excitation at the hippocampus enhances the *synaptic* response to a single input pulse.

Long-Term Potentiation (LTP)

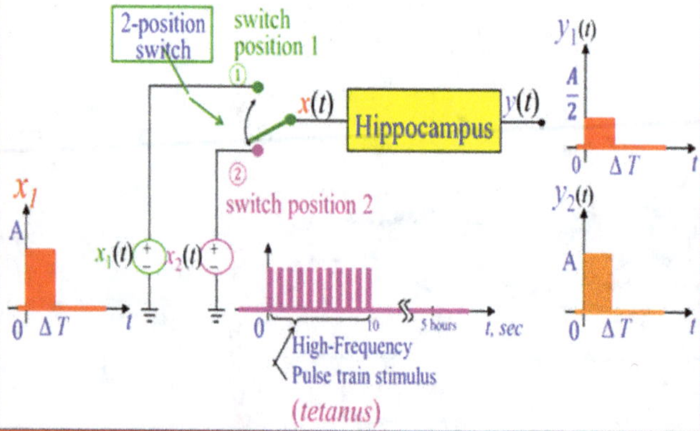

2-136

LTP depends only on Time-Integral of Pulse Train

Thinner pulses but with Larger amplitude produce the same effect

2-137

Long Term Potentiation

2-138

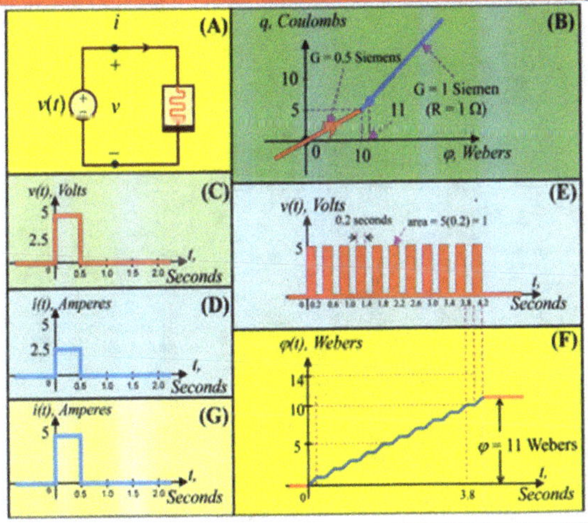

2-139

Ideal Electronic Synapses : Specifications

- Non-volatile

  No power supply is needed

- Analog

  Synaptic strength is tunable over a continuous range

- Plastic

  Synaptic strength can "tune" itself in response to dynamical history of the signal it receives

2-140

2-140

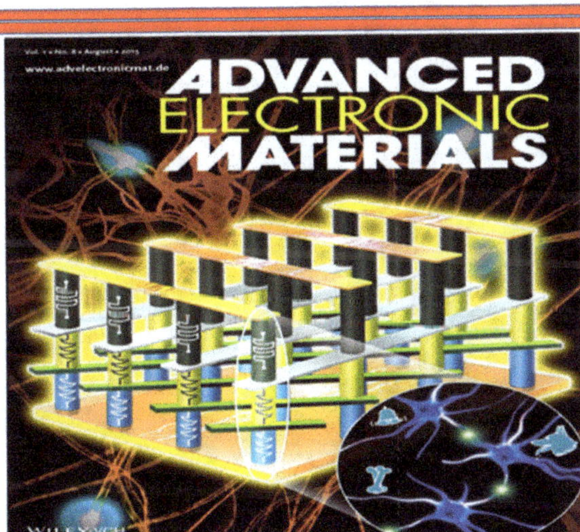

2-141

Pavlov's Classic Experiment Demonstrating Associative Learning 2-142

Ivan P. Pavlov

1904 Nobel Prize in Physiology or Medicine

Stimulus	Response
Unconditioned stimulus (UCS)	Unconditioned response (UCR)
Conditioned response (CS)	Conditioned response (CR)

Pavlov's Nobel-Prize Wining Experiment

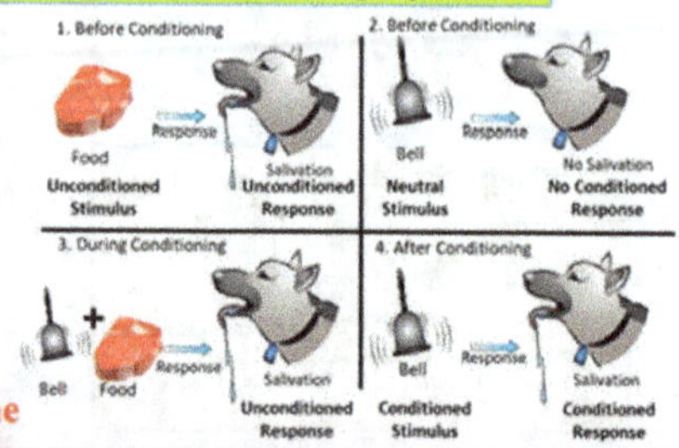

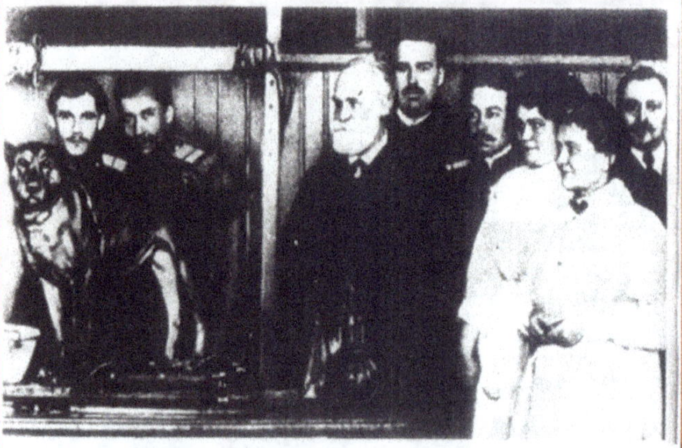

Pavlov assistants, visitors, and dog. Pavlovian (classical) conditioning is the simplest associative learning paradigm. 2-143

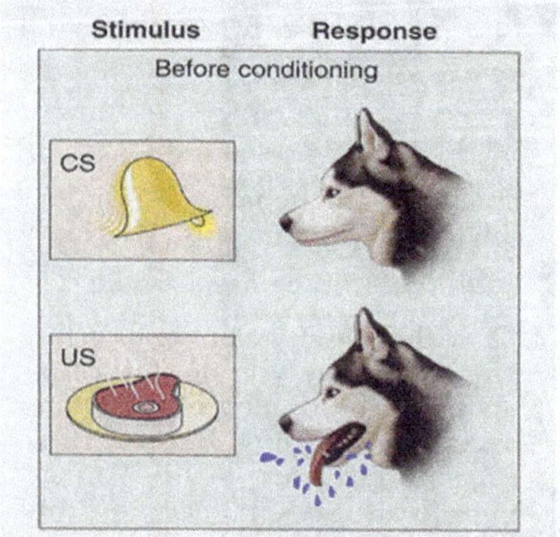

2-144

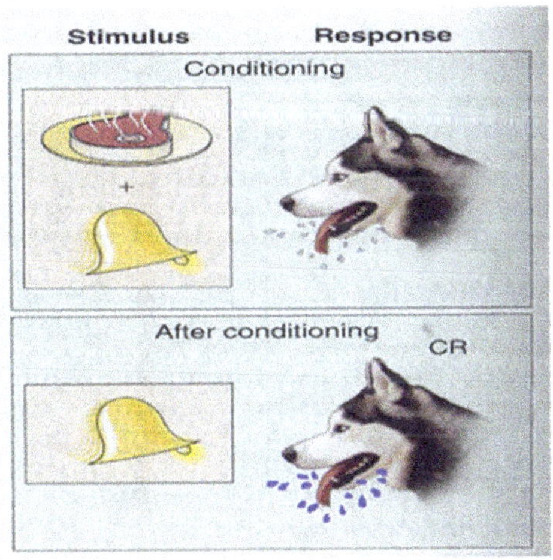

2-145

Associative Learning

In 350 BC,
Aristotle had already suggested that learning involves an association of ideas.

2-146

Associative Learning with Temporal Contiguity in a Memristive Circuit for Large-Scale Neuromorphic Networks

Yi Li, Lei Xu, Ying-Peng Zhong, Ya-Xiong Zhou, Shu-Jing Zhong, Yang-Zhi Hu, Leon O. Chua and Xian-Shui Miao.

Adv. Electronic Mater. 2015

2-147

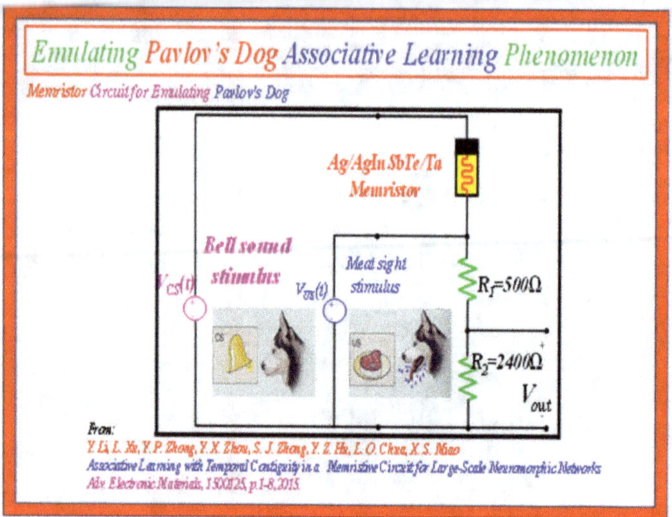

2-148

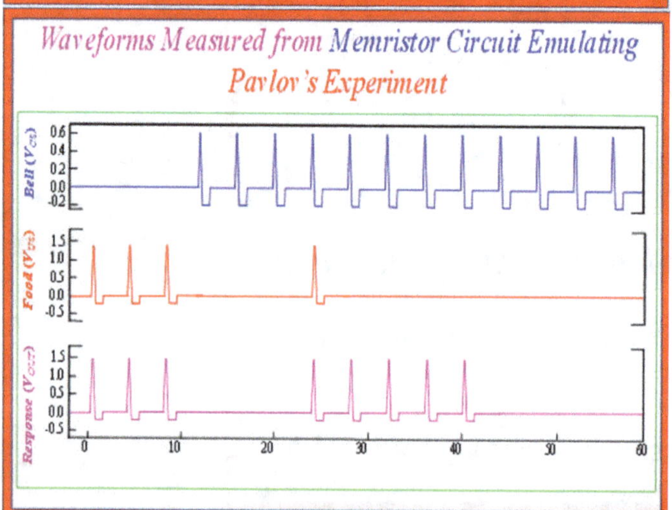

2-149

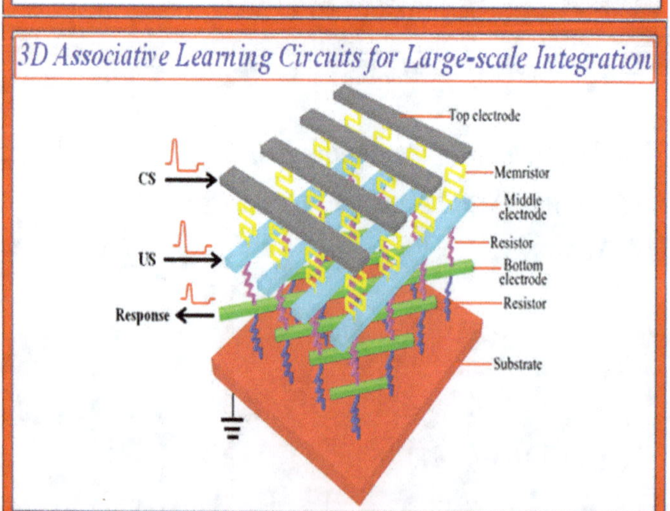

2-150

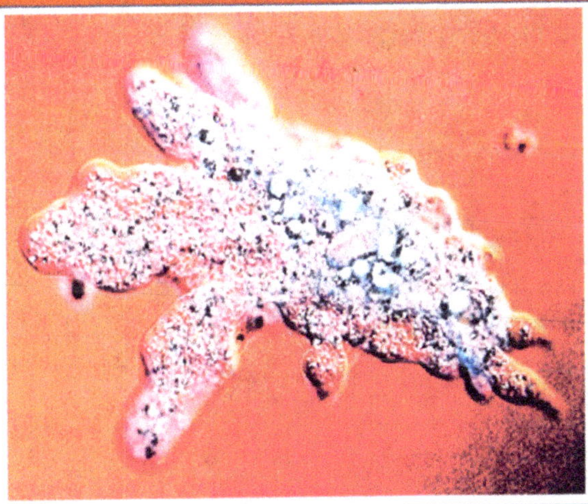

2-151

Amoeba

crawling speed :

1 cm/hour

at room temperature. 2-152

Amoeba

*does **not** have*

*a **brain** ,*

2-153

but it has

*a **mind** !*

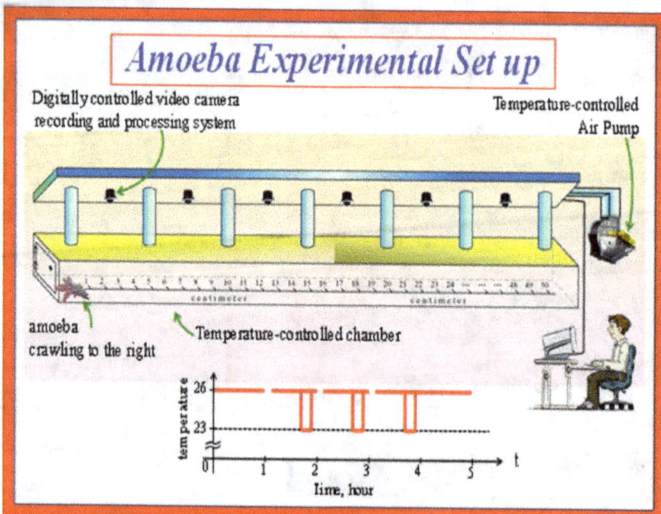

2-154

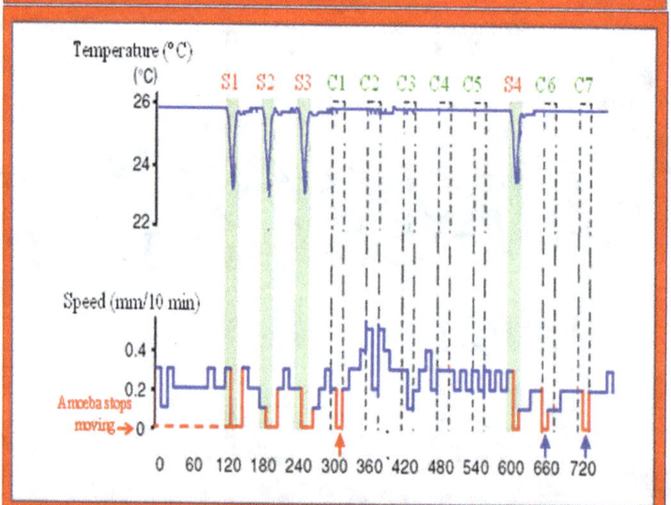

2-155

Amoeba Can Learn and Adapt !

When stimulated by a pattern of periodic

environmental changes,

an amoeba learns and changes

its behavior in anticipation of the

next stimulus to come.

2-156

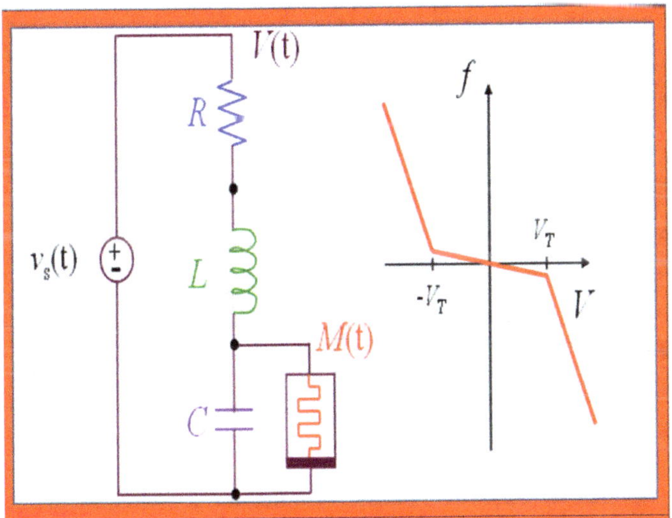

2-157

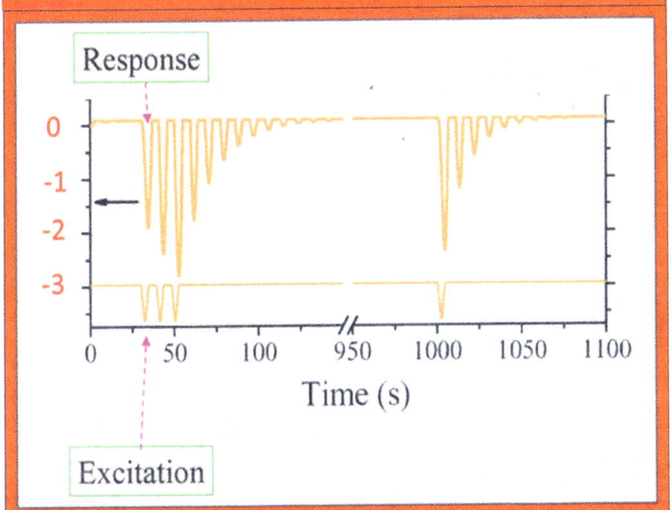

2-158

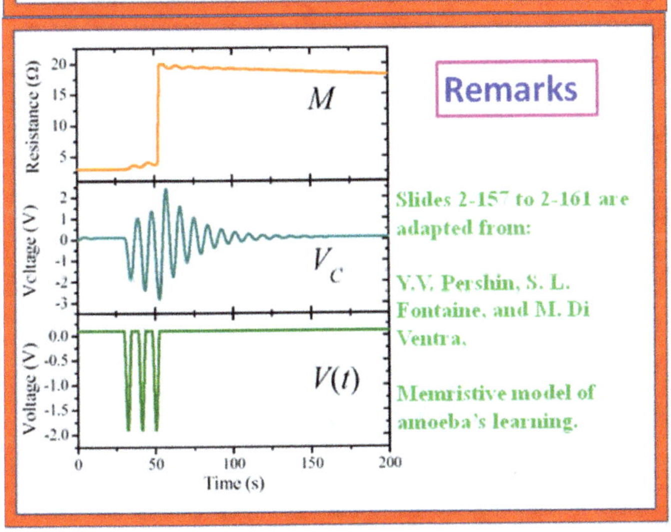

2-159

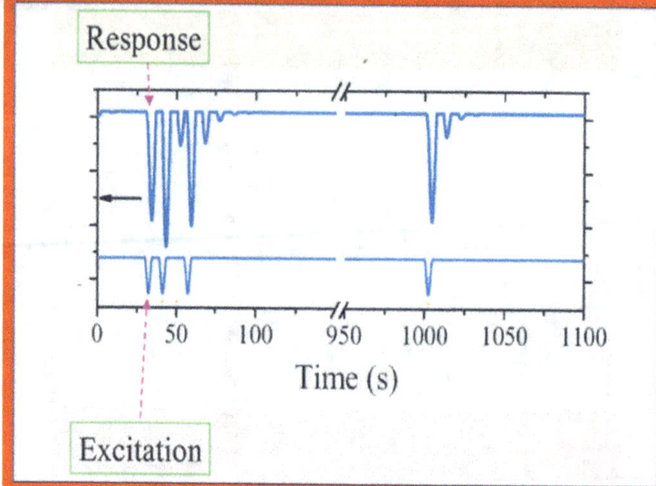

2-160

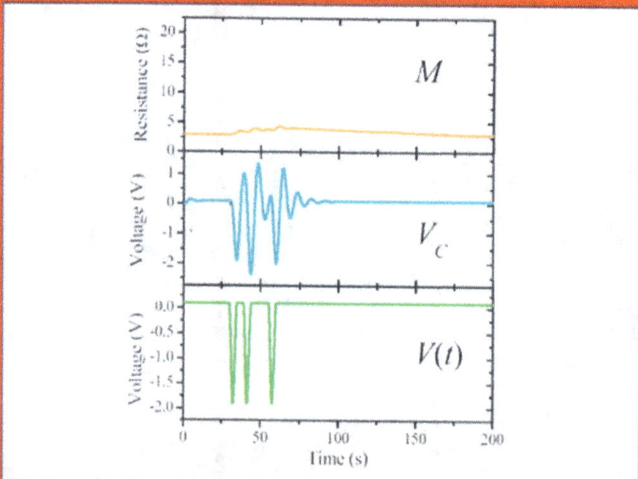

2-161

Are

Amoeba

intelligent ?

2-162

Dawn of Civilization

2-163

Amoeba is Smarter than early Human !

Discerning a ***periodicity*** is not easy, even for humans. When the ancient Egyptians recognized the regular ***periodicity*** of the flooding of the river Nile and succeeded in anticipating the next flood, this breakthrough triggered the invention of the ***calendar*** and was a symbol of the ***dawn of civilization***.

2-164

Synapses are Memristors!

2-165

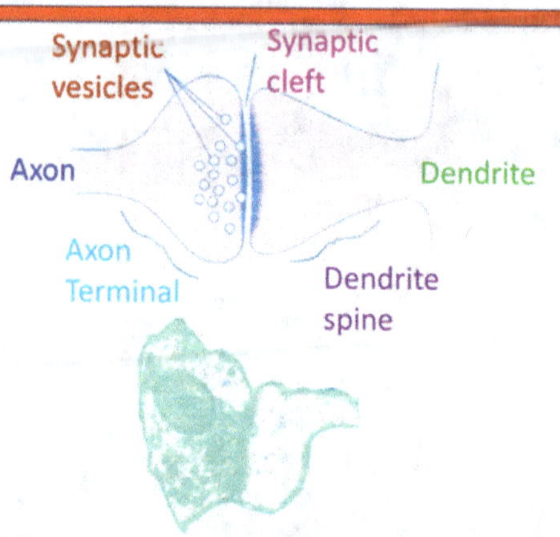

Parts of the synapses in a drawing (top) and in an actual electron microscope photo (bottom.) 2-166

There is a GAP !

The synapse between the axon and the dendrite is not physically connected ! 2-167

Synaptic Cleft

The width of the synaptic cleft is only about 200 – 300 Angstroms. 2-168

From Balloon to Neuron Topology

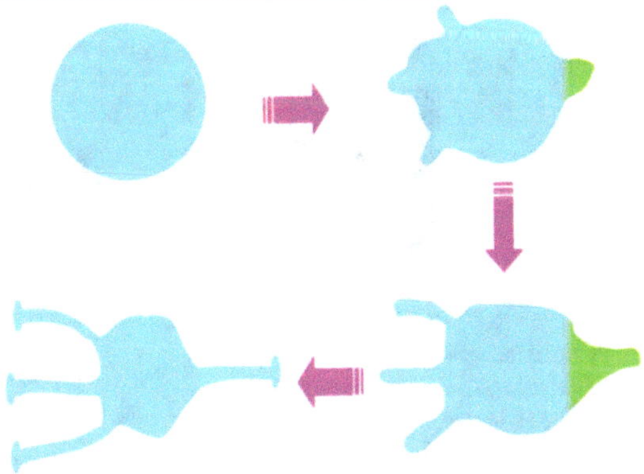

2-169

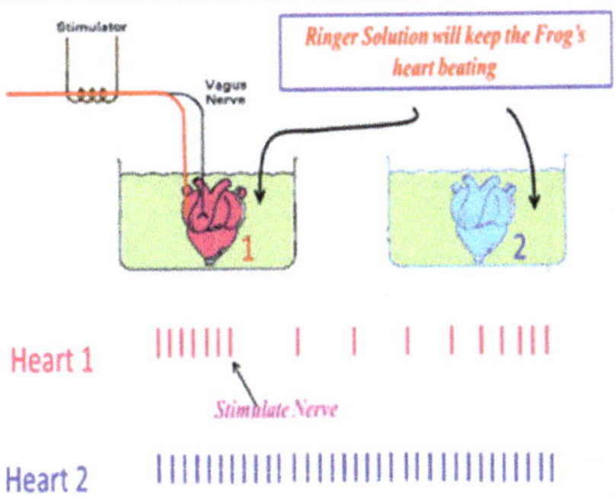

Otto Loewi

Nobel Prize in Medicine 1936.

Chemical transmission at a nerve-muscle synapse. In this famous experiment performed in 1921, Otto Loewi placed an innervated *(1)* and a non-innervated *(2)* heart in two separate chambers connected by a bridge of physiological saline. At the bottom is shown the rate of beating of both hearts before and after stimulation of the vagus nerve connected to the first heart.

2-170

Keeping Frog Heart Beating

*An isolated **frog heart** will continue to beat for a long time if it is immersed in a solution of **sodium**, **potassium**, and **calcium** of a specific proportion* 2-171

*(called **Ringer's solution** –discovered by **Sidney Ringer in 1881**)*

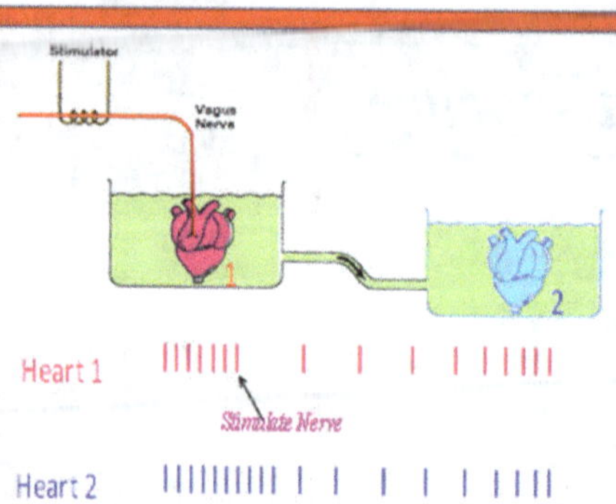

Otto Loewi

Nobel Prize in Medicine 1936. 2-172

Chemical transmission at a nerve-muscle synapse. In this famous experiment performed in 1921, Otto Loewi placed an innervated *(1)* and a non-innervated *(2)* heart in two separate chambers connected by a bridge of physiological saline. At the bottom is shown the rate of beating of both hearts before and after stimulation of the vagus nerve connected to the first heart.

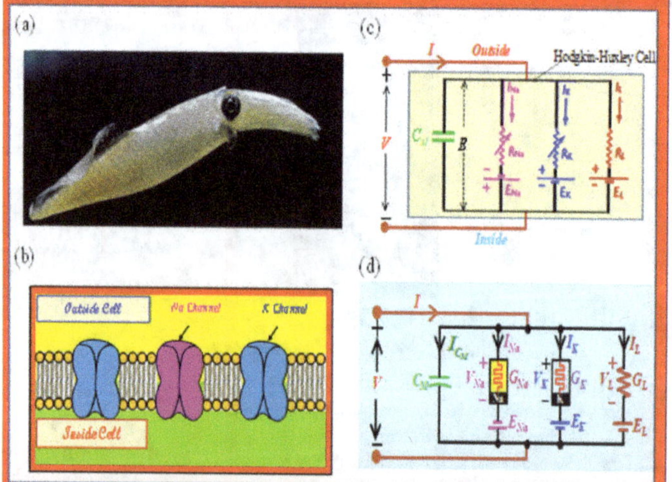

2-173

Ions and Ion Channels

Excitation and electrical signaling in the brain involve the movement of ions through ion channels. The Na^+, K^+, Ca^{2+}, and Cl^- ions are responsible for almost all of the transport dynamics. 2-174

Are Ion Channels Real !

Until 1964, no one knows whether the potassium and sodium ion channels postulated by Hodgkin and Huxley in their 1952 Nobel prize paper actually exist ! 2-175

Magic-Bullet Proof

The magic bullet which proved the potassium ion channel, and the sodium ion channel, are real comes from the FUGU ! 2-176

 2-177

The FUGU

The pufferfish is a very expensive delicacy offered only in high-end gourmet restaurants in Japan. 2-178

The *Japanese emperor* and his family are *prohibited from eating* the supreme Japanese delicacy called ***Fugu*** ! 2-179

It is also deadly because a small organ of the pufferfish contains a toxic molecule called tetrodotoxin, which is 1,000 times as powerful as cyanide. 2-180

The tetrodotoxin from the pufferfish is a tiny molecule which fits exactly the outer portion of the sodium channel's central pore, thereby plugging the opening completely, and suppressing the generation of any action potential. 2-181

What blocks Potassium Channels ?

Scorpion toxin blocks potassium channels by binding tightly to a site within the channel pore. Mackinnon used the toxin to identify the precise stretch of amino acids that forms the inside walls and selectivity filter of the channel for potassium ions. 2-182

2-183

Roderick MacKinnon

2003 Nobel Prize in Chemistry

Roderick MacKinnon's seminal paper was published in the May 2003 issue of Nature.

His discovery was so important and surprising that he was recognized with Nobel Prize only a few months later !

2-184

2-185

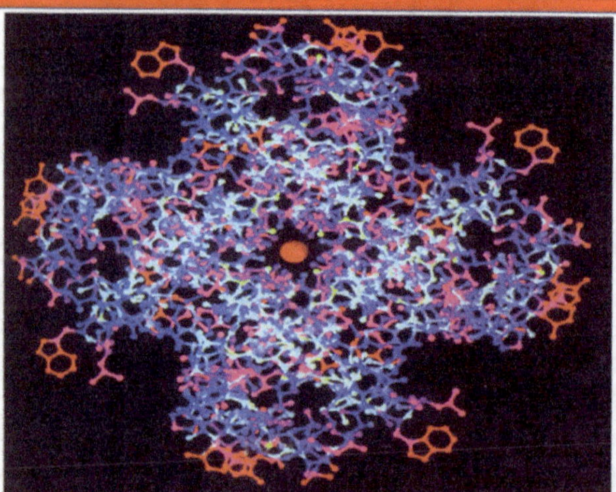

2-186

The preceding Picture is worth a Nobel Prize ! 2-187

Ion channels 2-188

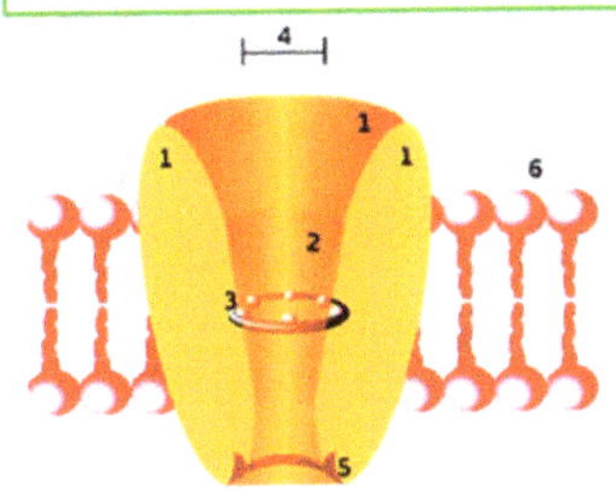

Ion channels are water-filled biological "*sub-nanotubes*" formed by large protein molecules.

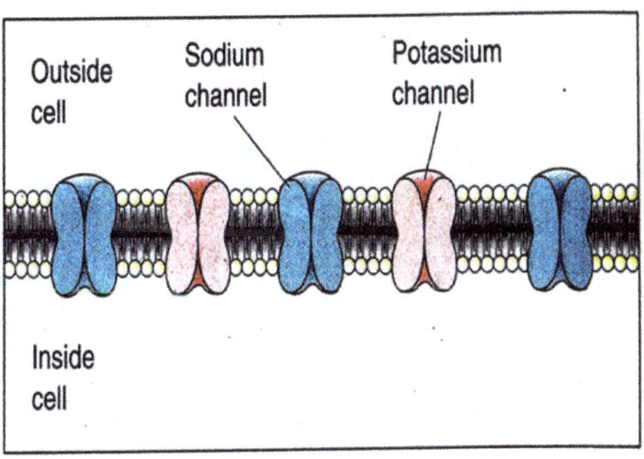

2-189

Ion Channels are like Bridges

An *ion channel* across a membrane is like a *bridge* across a river. 2-190

A *gated ion channel* across a membrane is like a *gated channel*, which can be open or close by a *drawbridge*.

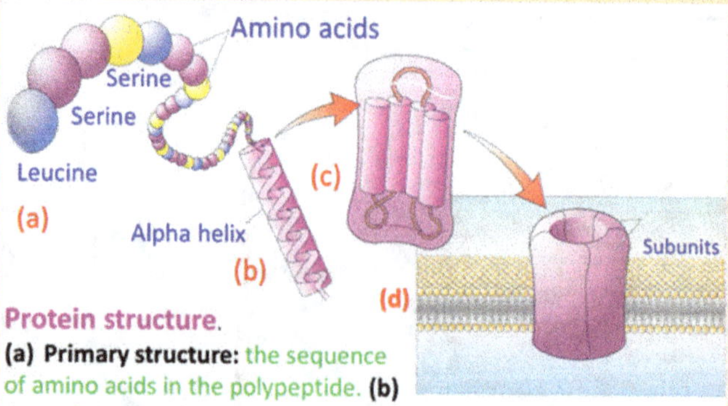

Protein structure. 2-191

(a) Primary structure: the sequence of amino acids in the polypeptide. **(b)** Secondary structure: coiling of a polypeptide into an alpha helix. **(c)** Tertiary structure: three-dimensional folding of a peptide. **(d)** Quaternary structure: different polypeptides bonded together to form a larger protein.

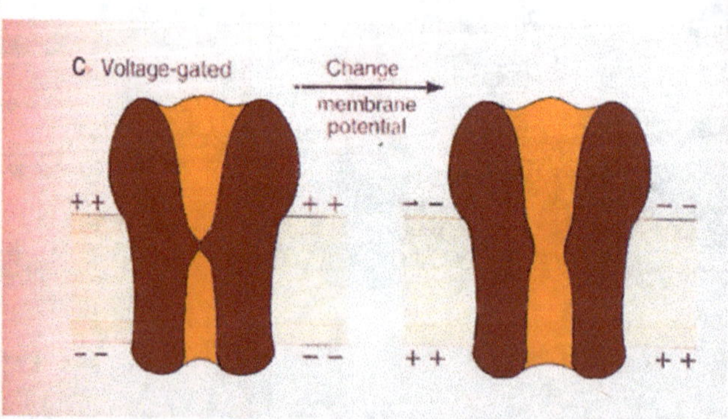

2-192

Voltage-Gated
Ion Channels
are
Memristor

2-193

Ion Channels are Memristor

Ion flows through an ion channel is simply current flows through a memristor.

2-194

Where there is smoke there is fire

2-195

Where there is
ion 2-196
there is
memristor

Lecture 3

10 Things You 2-197
Didn't Know about
Memristors

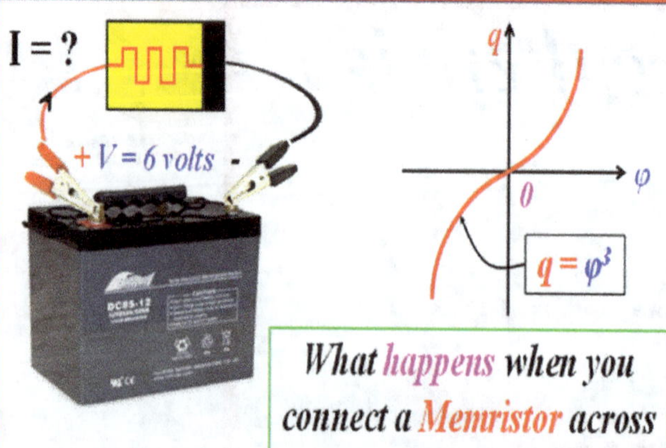 2-198

What happens when you connect a Memristor across a battery ?

Warm up Reads

1. S. P. Adhikari, M. P. Sah, H. Kim & L. O. Chua, "Three fingerprints of memristors," IEEE Trans. Circuit and System-I, 60, 3008-3021, 2013.
2. L. Chua "Resistance Switching Memories Are Memristors," Applied Physics A, 102, pp.765-783, 2011.
3. L. Chua, "If It's Pinched It's a Memristor," Semicond. Sci. Technol., 29, pp.104001(42pp), 2014.
4. L. Chua, "Everything You Wish to Know About Memristor But Are Afraid to Ask," RADIOENGINEERING, vol.24, pp.319-368, 2015.
5. L. Chua, "How we predicted the memristor," Nature Electronics, vol. 1, 2018, pp.322.

2-199

"A model is a mathematical construction, which supplemented with some verbal explanation, describes the observed phenomena. Such a mathematical construction is proved if and only if it works, that is it describes precisely a wide range of phenomena. Furthermore it has to satisfy certain aesthetic criteria, i.e. it has to be more or less simple compared to the described phenomena".

J. Von Neumann

2-200

End Quote for Lecture 2

A model must be simple,
but not simpler.

Albert Einstein

2-201

Appendix 2-202

2-203

2-204

Déjà vu Sexual Response ?

Attracted to a voluptuous *female*, a *male rat* will have sex with her six or seven times in a period of one or two hours. After the last copulation he appears exhausted and becomes inactive for 30 minutes or longer. 2-205

 2-206

Sexual Déjà vu !

This is not *fatigue*. The apparently exhausted male rat will instantly resume his passionate sexual repertoire if he encounters moments later a *new* seductive *female* ! 2-207

Sexual Déjà vu is Reciprocal

When a new ***male rat*** is introduced instead, it is the ***female rat*** that springs into action.

Remarks: Slides 205-208 are adapted from "**Memory**" by L.R. Squire and E.R. Kandel

2-208

2-209

2-210

Lecture 3

10 Things You Didn't Know about Memristors 2-211

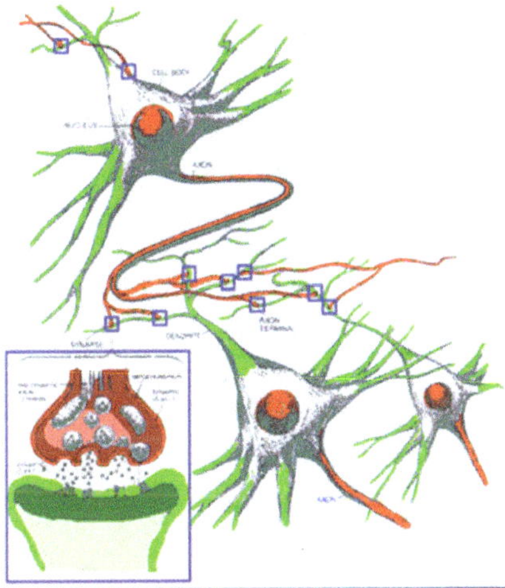 2-212

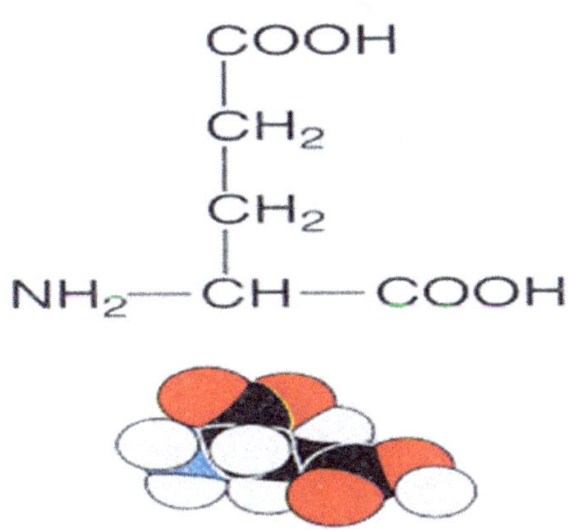 2-213

Glutamate

Glutamate is the key neurotransmitter responsible for LTP memory. 2-214

Glutamate: A Common Excitatory Neurotransmitter

When released at a *pre-synaptic* membrane, some *glutamate* molecules *diffuse* across the *synaptic cleft* (in about 5 microseconds) and bind into a specialized *glutamate receptor protein* embedded in the *postsynaptic* membrane. 2-215

This is a very complex *molecular machine* made of 4 similar parts that join together to form a *doughnut-shaped structure* around a *central pore*. In this resting state, this *pore* is shut tight, but when *glutamate* binds this receptor, the *gate* to the pore *opens*, allowing both *sodium* and *potassium* ions to flow through the "*ion channel*".

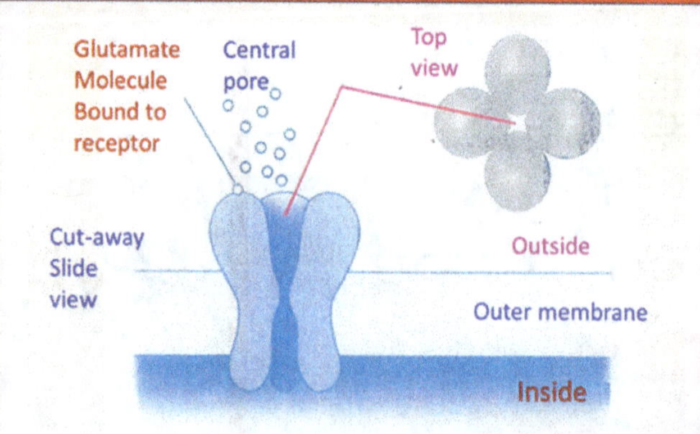

Schematic drawing of a glutamate receptor in the postsynaptic mem-brane. Glutamate binding to its receptor opens the central pore, the ion channel. 2-216

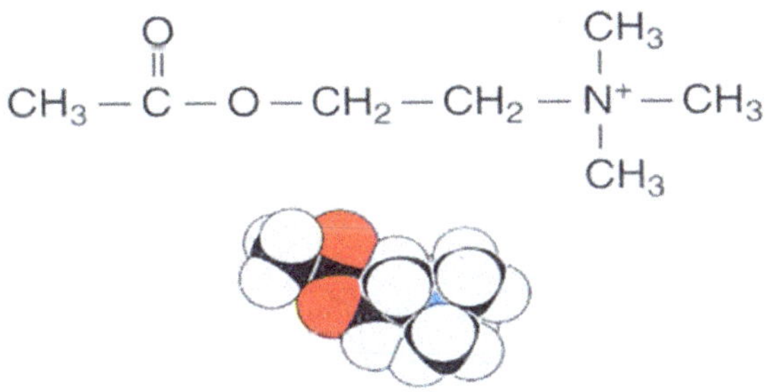

Acetylcholine 2-217

Acetylcholine : A Common Fast Acting Excitatory Neurotransmitter

Acetylcholine acts like *Glumatate* by opening an *ion channel* that lets both *Sodium* rush in and *Potassium* out. 2-218

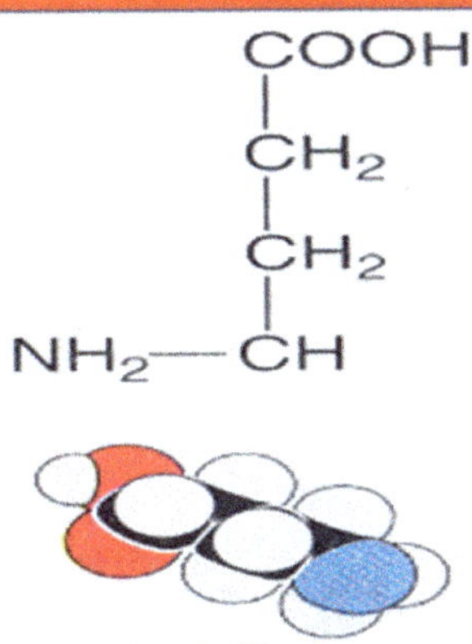

GABA
(Gamma-aminobutyric acid) 2-219

GABA: A Common Fast-Acting Inhibitory Neurotransmitter

GABA binds to ***receptors*** which *opens* a channel that let ***chloride*** ions to flow into the ***postsynaptic neuron***. Chloride ions Cl^- have a ***negative*** charge (-1), and thus make the membrane potential ***more negative*** (within a few milliseconds), thereby ***inhibiting*** (rather than stimulating) the "***firing***" of ***spikes*** (***action potentials***). 2-220

Valium belongs to a family of drugs that bind with GABA, and neurons can become so inhibited you feel dozy. 2-221

NH_2-CH_2-COOH

Glycine 2-222

Glycine: A Common Fast-Acting Inhibitory Neurotransmitter

When a Glycine molecule binds to a receptor, it acts like GABA by making the membrane potential to become more negative (within a few milliseconds), thereby inhibiting the "firing" of spikes (action potentials). 2-223

How the Poison Strychnine kills ?

The poison strychnine, which figures prominently in mystery novels, blocks glycine receptors, and prevents their activation. 2-224

NATURE ELECTRONICS | VOL 1 | MAY 2018 | 322 | www.nature.com/natureelectronics

How we predicted the memristor

The memristor — a resistor with memory — was first postulated back in 1971, but it took nearly four decades before it was experimentally confirmed. Leon O. Chua explains how he developed a nonlinear circuit theory in which the memristor emerges naturally as the fourth basic circuit element.

Leon O. Chua 2-225

The dawn of nonlinear electronics was ushered in by newly minted two-terminal electronic devices bearing such intimidating monikers as the Esaki diode, Josephson junction, varactor, thyristor, the IMPATT (impact ionization avalanche transit-time) diode and the Ovonic threshold switch. The proliferation of these exotic devices in the 1960s was an exciting period for electronic engineers. At this time, I was working on my PhD, trying to make sense of these strongly nonlinear and dynamical electronic devices.

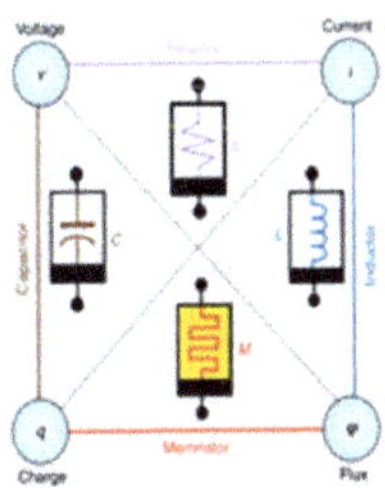

circuit elements, namely, the resistor (defined by a relationship between v and i), the inductor (defined by a relationship between φ and i), and the capacitor (defined by a relationship between q and v). Only one relationship remains undefined, the relationship between φ and q. From the logical as well as axiomatic points of view, it is necessary for the sake of completeness to postulate the existence of a fourth basic two-terminal circuit element which is characterized by a φ–q relationship. This element will henceforth be called the memristor because [...] it behaves somewhat like a nonlinear resistor with memory."

Lecture 3

10 Things You Didn't Know About Memristors

Lecture 3

10 Things You Didn't Know about Memristors

3-1

Duality Principle in Circuit Theory

For every variable, element, definition, concept, theorem,...., etc,
there exist a dual variable, element, definition, concept, theorem,...., etc.

3-2

Examples

Voltage v ↔ Current i

Flux φ ↔ Charge q

Voltage source ↔ Current source

Voltage - controlled ↔ Current - controlled

Resistance (Ω) ↔ Conductance (S)

Capacitance (C) ↔ Inductance (L)

Series connection ↔ Parallel connection

Charge-controlled memristor ↔ Flux-controlled memristor

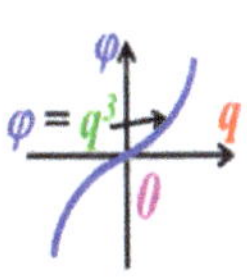

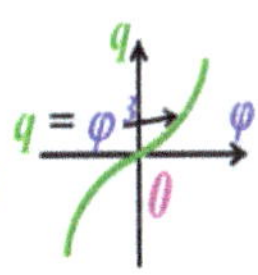

4 Basic Circuit Variables

3-3

- Voltage $v(t)$
- Current $i(t)$
- Charge $q(t) \triangleq \int_{-\infty}^{t} i(\tau)\,d\tau$, or $i = \dfrac{dq(t)}{dt}$
- Flux $\varphi(t) \triangleq \int_{-\infty}^{t} v(\tau)\,d\tau$, or $v = \dfrac{d\varphi(t)}{dt}$

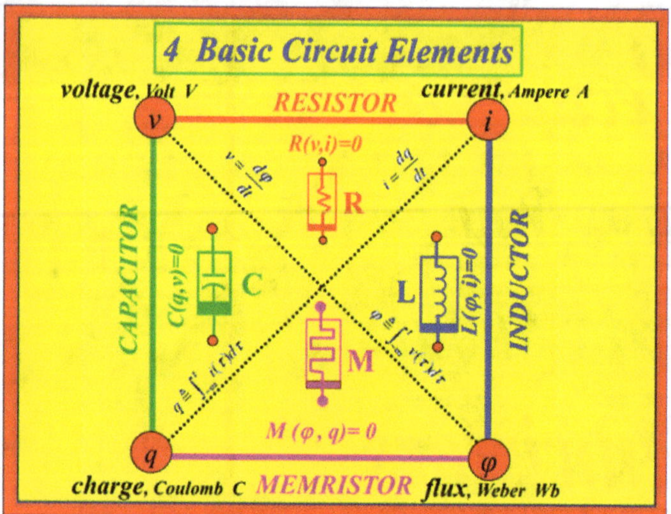
4 Basic Circuit Elements
voltage, Volt V
current, Ampere A
RESISTOR
R(v,i)=0
R
CAPACITOR
C(q,v)=0
C
INDUCTOR
L(φ,i)=0
L
M
M (φ, q)= 0
charge, Coulomb C
MEMRISTOR
flux, Weber Wb

3-4

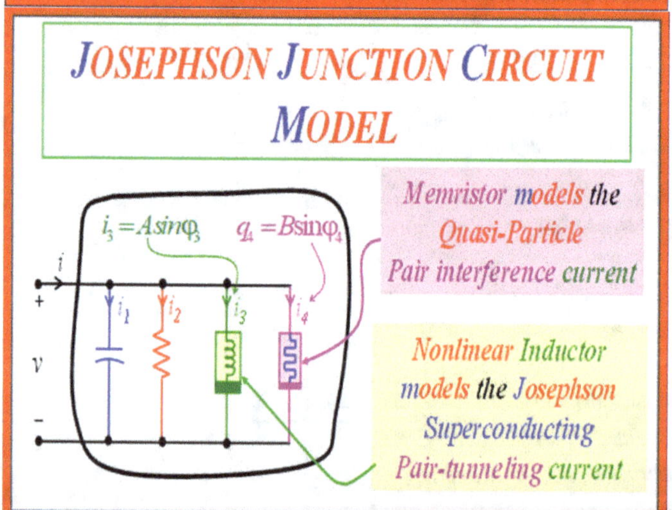
JOSEPHSON JUNCTION CIRCUIT MODEL
Memristor models the Quasi-Particle Pair interference current
Nonlinear Inductor models the Josephson Superconducting Pair-tunneling current

3-5

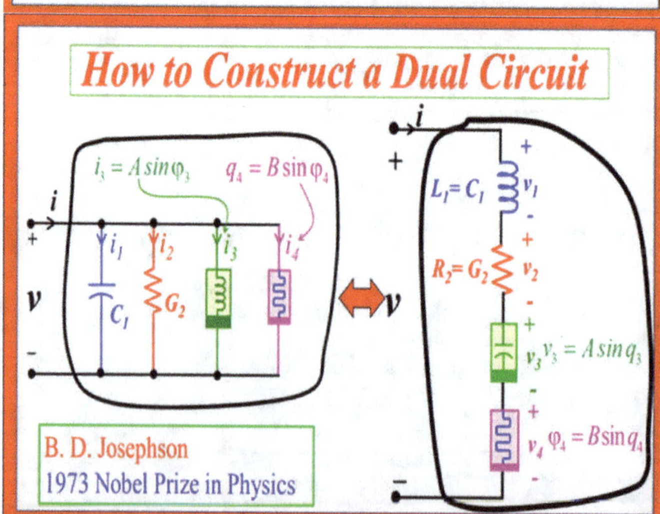
How to Construct a Dual Circuit
B. D. Josephson
1973 Nobel Prize in Physics

3-6

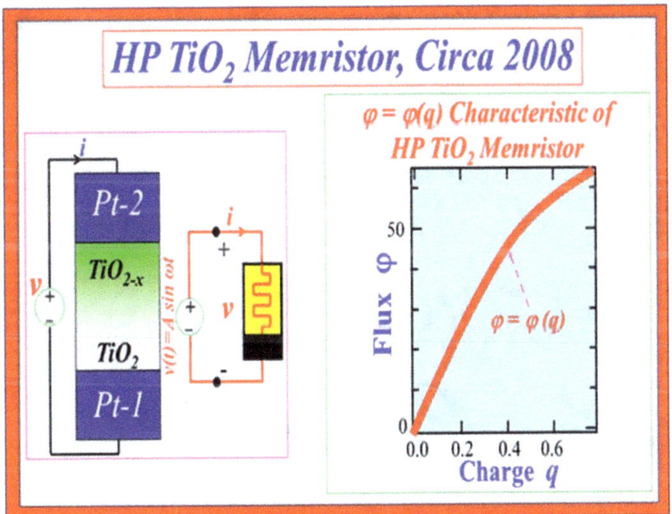
HP TiO$_2$ Memristor, Circa 2008
i
Pt-2
TiO_{2-x}
TiO_2
Pt-1
v
+
-
$v(t) = A \sin \omega t$
i
+
v
-
$\varphi = \varphi(q)$ Characteristic of HP TiO$_2$ Memristor
50
0
Flux φ
$\varphi = \varphi(q)$
0.0 0.2 0.4 0.6
Charge q

3-7

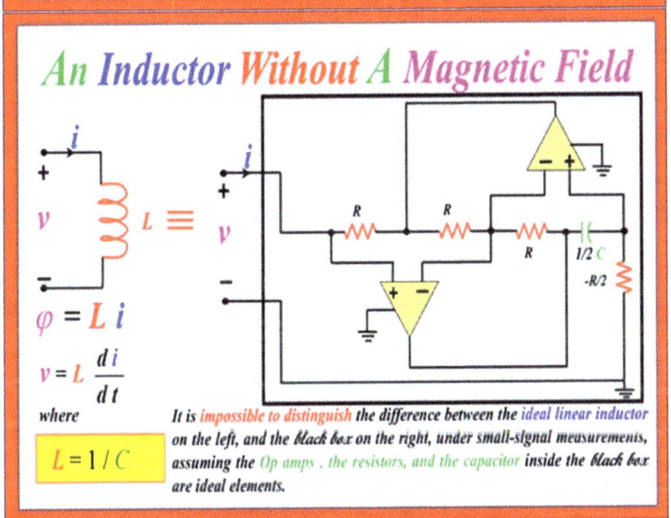
An Inductor Without A Magnetic Field
i
+
v
-
L ≡
$\varphi = L\, i$
$v = L \frac{d\,i}{d\,t}$
where
$L = 1/C$
R
R
R
1/2 C
-R/2
It is impossible to distinguish the difference between the ideal linear inductor on the left, and the black box on the right, under small-signal measurements, assuming the Op amps, the resistors, and the capacitor inside the black box are ideal elements.

3-8

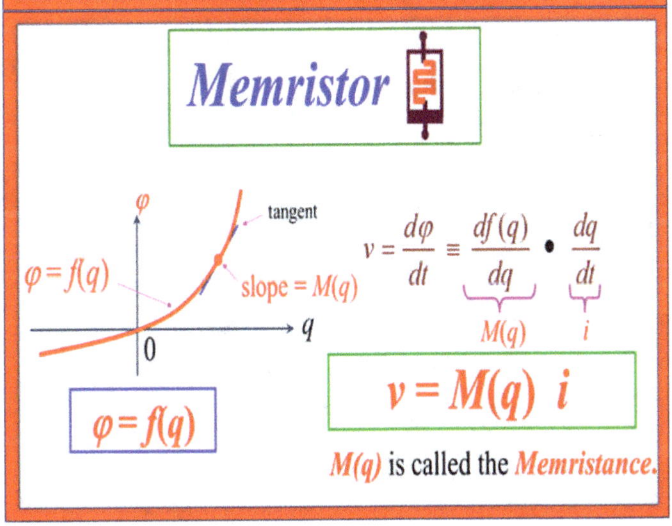
Memristor
φ
tangent
$\varphi = f(q)$
slope = $M(q)$
q
0
$v = \frac{d\varphi}{dt} \equiv \frac{df(q)}{dq} \bullet \frac{dq}{dt}$
$M(q)$
i
$\varphi = f(q)$
$v = M(q)\; i$
$M(q)$ is called the Memristance.

3-9

Memristor is defined by a State - Dependent Ohm's Law

3-10

How To Recover Memristor Constitutive Relation $\varphi = \hat{\varphi}(q)$ From Its Memristance $M(q)$?

$$v = M(q)\, i$$

$$\dot{q} = i$$

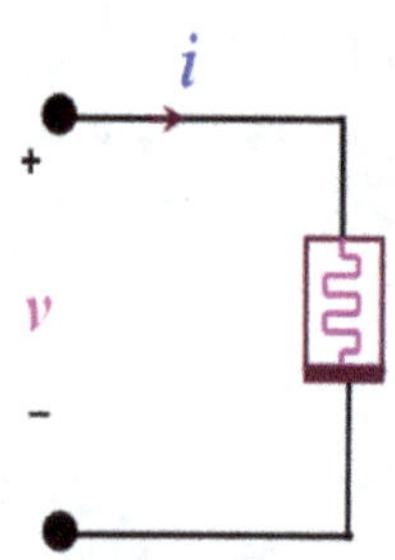

Constitutive Relation :

$$\varphi = \int M(q)\, dq \triangleq \hat{\varphi}(q)$$

where $\hat{\varphi}(q)$ is unique to within an arbitrary constant.

3-11

How To Recover The Constitutive relation of An Ideal Memristor from its Memristance ?

Given the memristance $R(q)$

We can recover its associated Constitutive relation via

$$\varphi = \hat{\varphi}(q) = \varphi_0 + \int R(q)\, dq$$

where $\varphi_0 = \hat{\varphi}(0)$

Example :

Given Memristance

$R(q) = 1 + q^2$

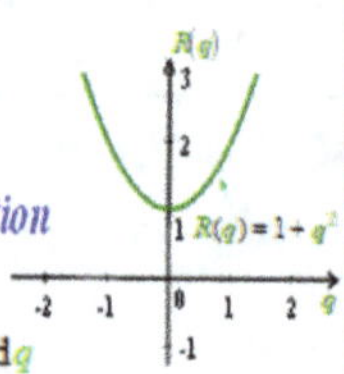

Constitutive relation

$$\begin{aligned}\varphi &= \hat{\varphi}(q) \\ &= \varphi_0 + \int (1+q^2)\, dq \\ &= \varphi_0 + q + \frac{1}{3}q^3 \\ &= q + \frac{1}{3}q^3\end{aligned}$$

assuming,

$\varphi(0) = 0$

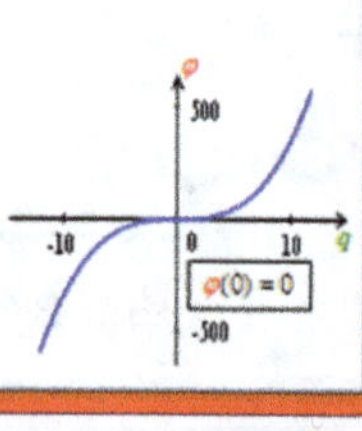

3-12

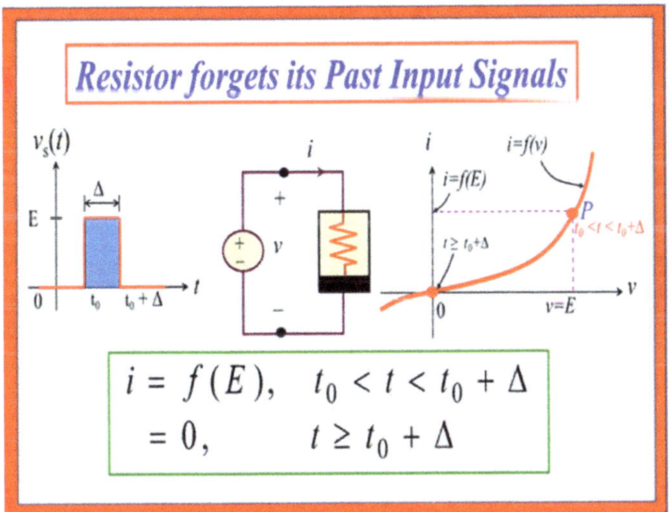
Resistor forgets its Past Input Signals
$v_s(t)$
Δ
E
0
t_0
$t_0 + \Delta$
t
i
+
v
−
i
$i=f(v)$
$i=f(E)$
P
$t_0 < t < t_0+\Delta$
$t \geq t_0+\Delta$
0
$v=E$
v
$i = f(E), \quad t_0 < t < t_0 + \Delta$
$= 0, \quad t \geq t_0 + \Delta$

3-13

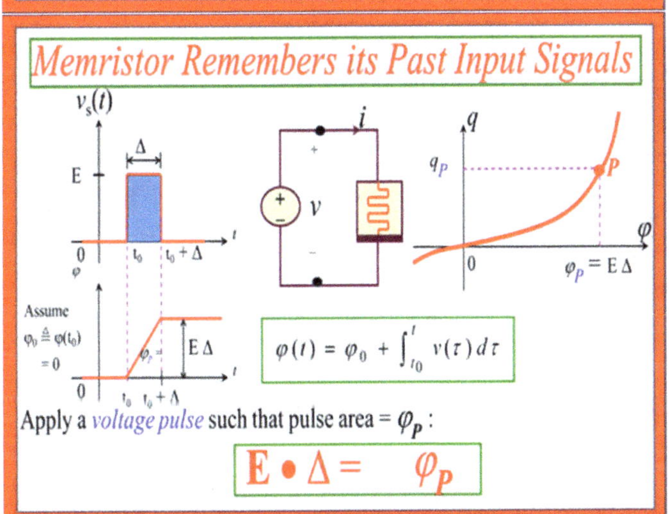
Memristor Remembers its Past Input Signals
$v_s(t)$
Δ
E
0
t_0
$t_0 + \Delta$
t
φ
Assume
$\varphi_0 \triangleq \varphi(t_0) = 0$
E Δ
i
+
v
−
q
q_P
P
φ
0
$\varphi_P = E\Delta$
$\varphi(t) = \varphi_0 + \int_{t_0}^{t} v(\tau)\,d\tau$
Apply a voltage pulse such that pulse area = φ_P :
$E \bullet \Delta = \varphi_P$

3-14

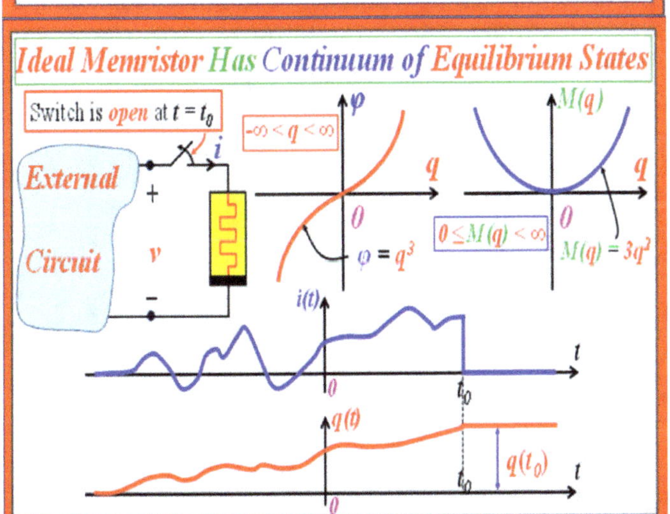
Ideal Memristor Has Continuum of Equilibrium States
Switch is open at $t = t_0$
External
Circuit
i
+
v
−
$-\infty < q < \infty$
φ
q
0
$\varphi = q^3$
M(q)
q
0
$0 \leq M(q) < \infty$
$M(q) = 3q^2$
i(t)
t
0
t_0
q(t)
$q(t_0)$
t
t_0
0

3-15

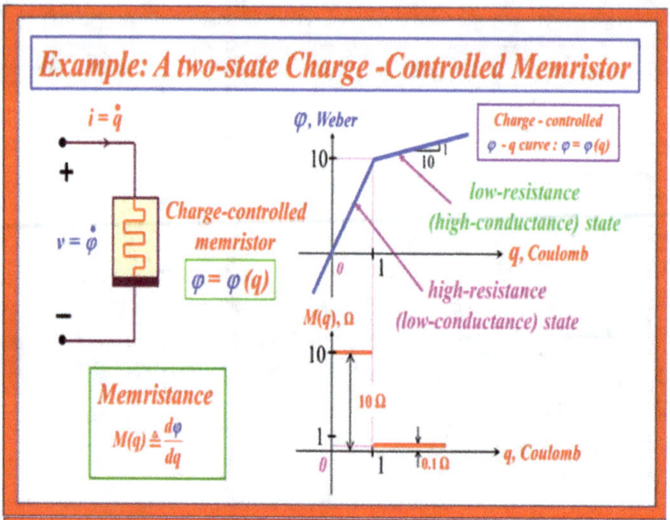
Example: A two-state Charge -Controlled Memristor
$i=\dot{q}$
$v=\dot{\varphi}$
Charge-controlled memristor
$\varphi=\varphi(q)$
Memristance
$M(q)\triangleq\frac{d\varphi}{dq}$
φ, Weber
Charge - controlled φ - q curve : φ = φ(q)
low-resistance (high-conductance) state
high-resistance (low-conductance) state
q, Coulomb
M(q), Ω
10 Ω
0.1 Ω
q, Coulomb

3-16

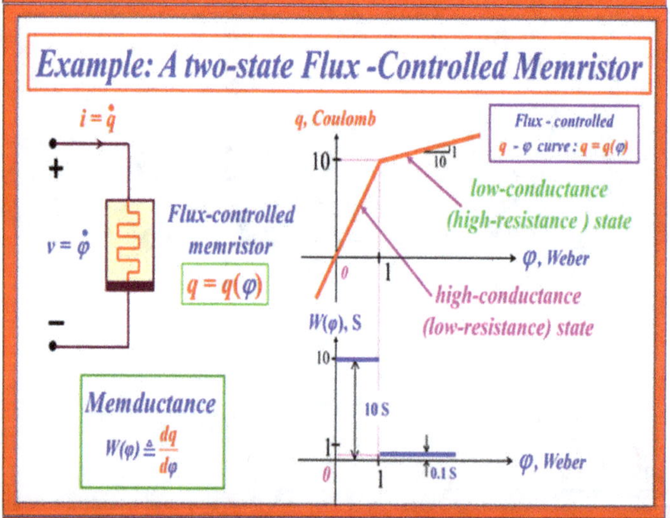
Example: A two-state Flux -Controlled Memristor
$i=\dot{q}$
$v=\dot{\varphi}$
Flux-controlled memristor
$q=q(\varphi)$
Memductance
$W(\varphi)\triangleq\frac{dq}{d\varphi}$
q, Coulomb
Flux - controlled q - φ curve : q = q(φ)
low-conductance (high-resistance) state
high-conductance (low-resistance) state
φ, Weber
W(φ), S
10 S
0.1 S
φ, Weber

3-17

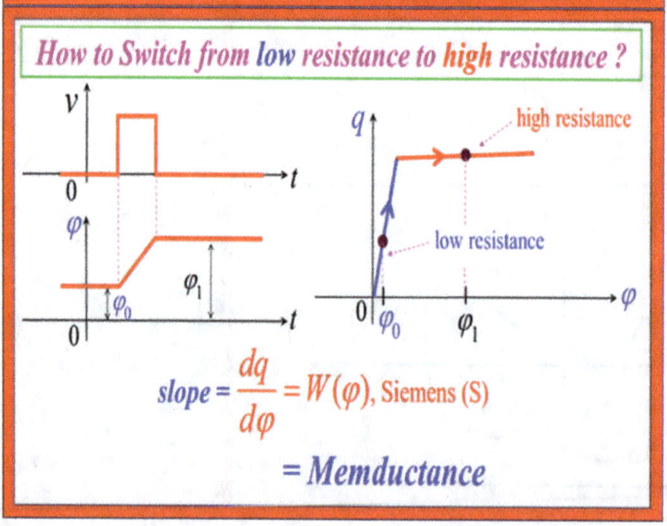
How to Switch from low resistance to high resistance ?
high resistance
low resistance
$\text{slope}=\frac{dq}{d\varphi}=W(\varphi)$, Siemens (S)
= Memductance

3-18

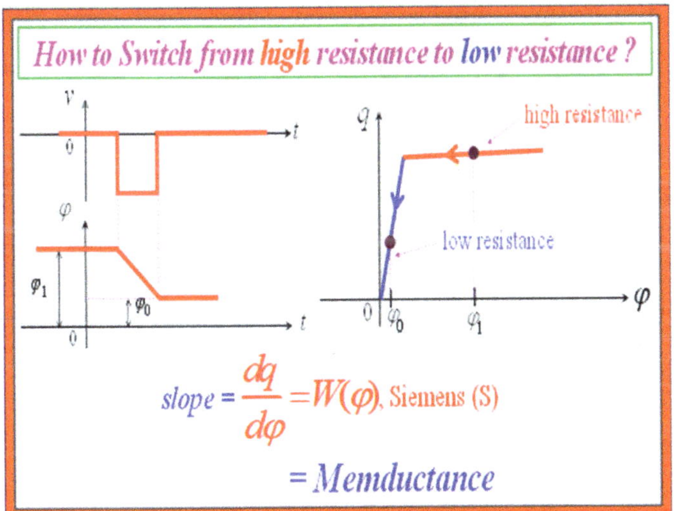
How to Switch from high resistance to low resistance ?
v
t
0
φ
φ_1
φ_0
0
t
q
high resistance
low resistance
0
φ_0
φ_1
φ
$slope = \frac{dq}{d\varphi} = W(\varphi)$, Siemens (S)
= Memductance

3-19

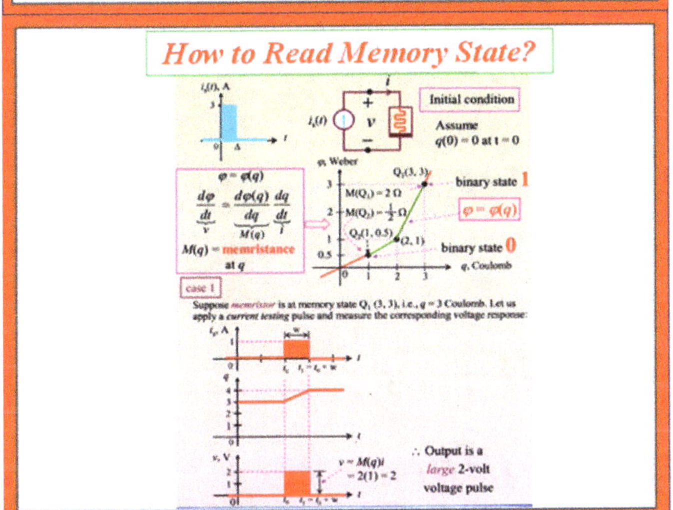
How to Read Memory State?
Initial condition
Assume
q(0) = 0 at t = 0
φ, Weber
$Q_1(3, 3)$
binary state 1
$M(Q_1) = 2\ \Omega$
$M(Q_2) = \frac{1}{2}\ \Omega$
$\varphi = \varphi(q)$
$Q_2(1, 0.5)$
(2, 1)
binary state 0
q, Coulomb
M(q) = memristance at q
case 1
Suppose memristor is at memory state Q_1 (3, 3), i.e., q = 3 Coulomb. Let us apply a current testing pulse and measure the corresponding voltage response:
v = M(q)i = 2(1) = 2
∴ Output is a large 2-volt voltage pulse

3-20

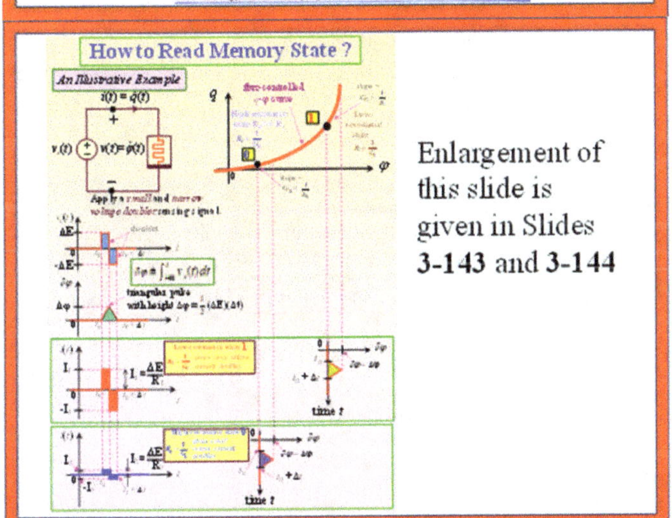
How to Read Memory State ?
An Illustrative Example
triangular pulse
time t
Enlargement of this slide is given in Slides 3-143 and 3-144

3-21

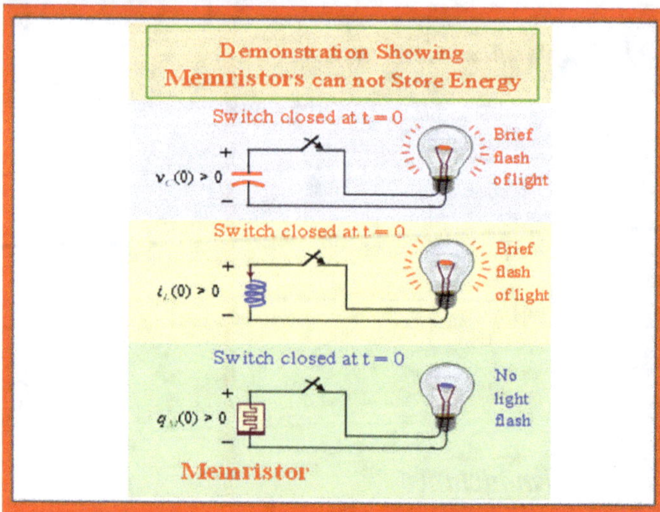

3-22

Memristor does not store energy

Memristor stores Information

3-23

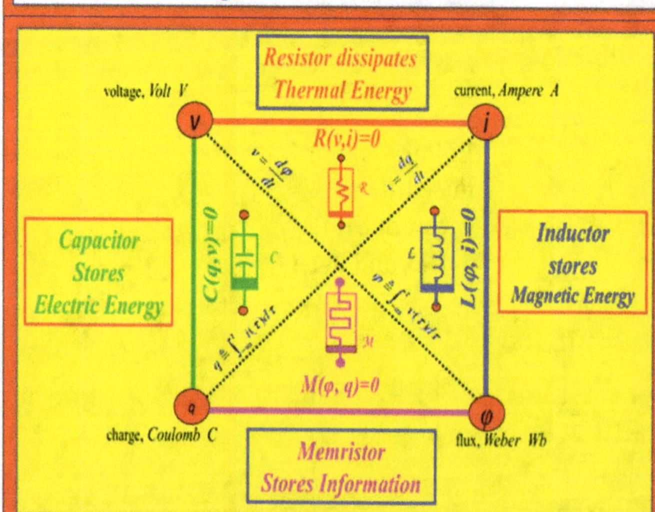

3-24

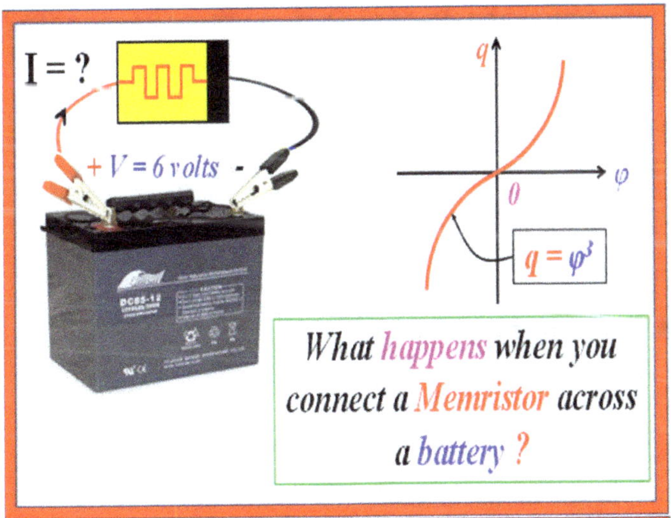
I = ?
+ V = 6 volts -
q
φ
0
q = φ³
What happens when you connect a Memristor across a battery ?

3-25

My Memristor is Melting !
VISbi 2014

3-26

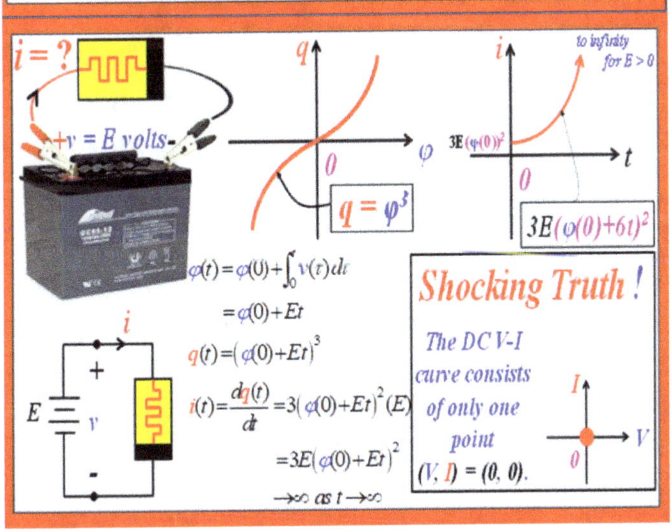
i = ?
+v = E volts-
q
φ
0
q = φ³
i
to infinity for E > 0
3E(φ(0))²
t
0
3E(φ(0)+6t)²
φ(t) = φ(0) + ∫₀ v(τ)dτ
= φ(0) + Et
q(t) = (φ(0)+Et)³
i(t) = dq(t)/dt = 3(φ(0)+Et)²(E)
= 3E(φ(0)+Et)²
→∞ as t→∞
i
+
E
v
-
Shocking Truth !
The DC V-I curve consists of only one point (V, I) = (0, 0).
I
V
0

3-27

The

Ideal Memristor

does not have a

DC V-I Curve ! 3-28

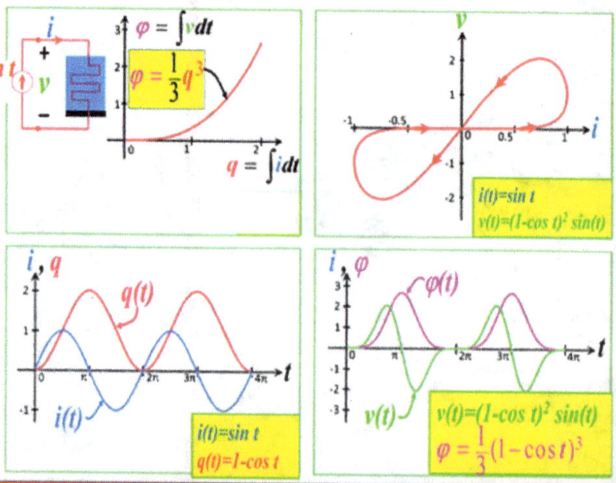

3-29

Pinched Hysteresis Loop

Any hysteresis loop which passes through the origin is called a Pinched Hysteresis Loop

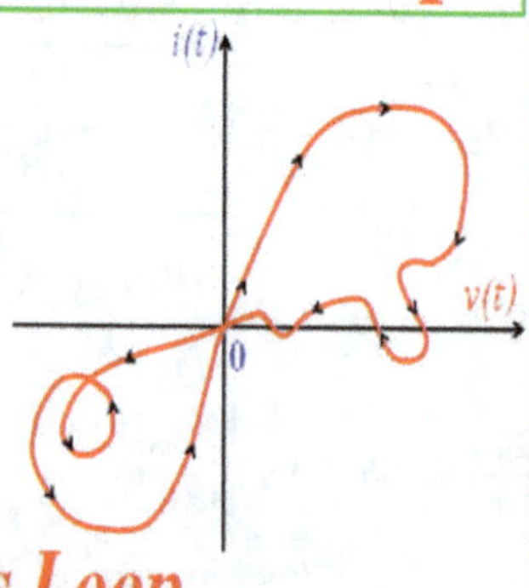

3-30

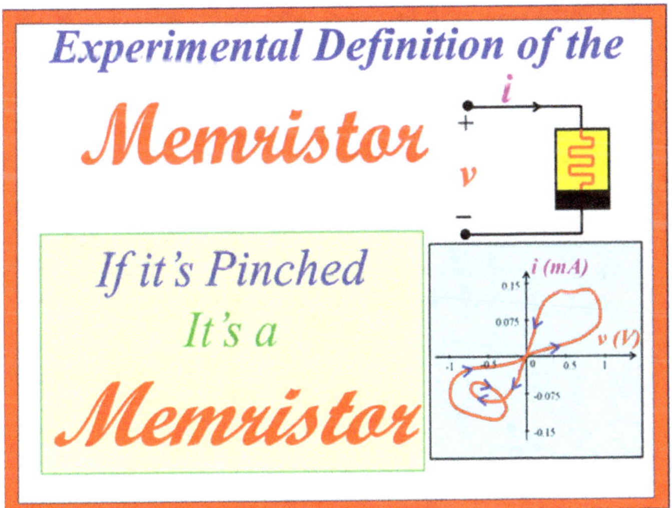
Experimental Definition of the
Memristor
i
+
v
−
If it's Pinched
It's a
Memristor
i (mA)
v (V)

3-31

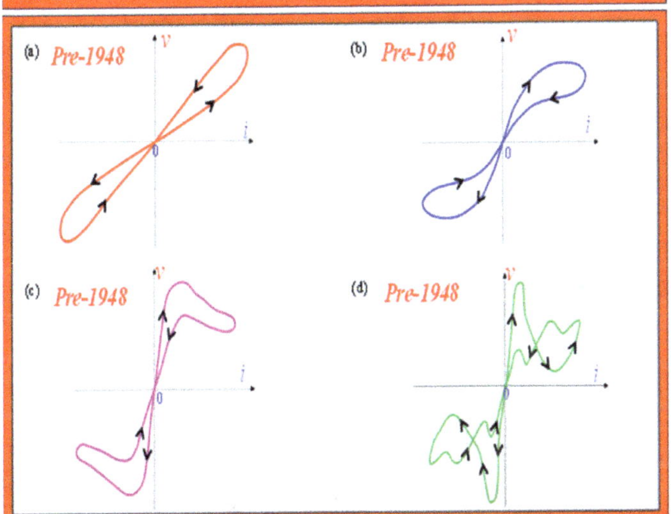
(a) Pre-1948
(b) Pre-1948
(c) Pre-1948
(d) Pre-1948

3-32

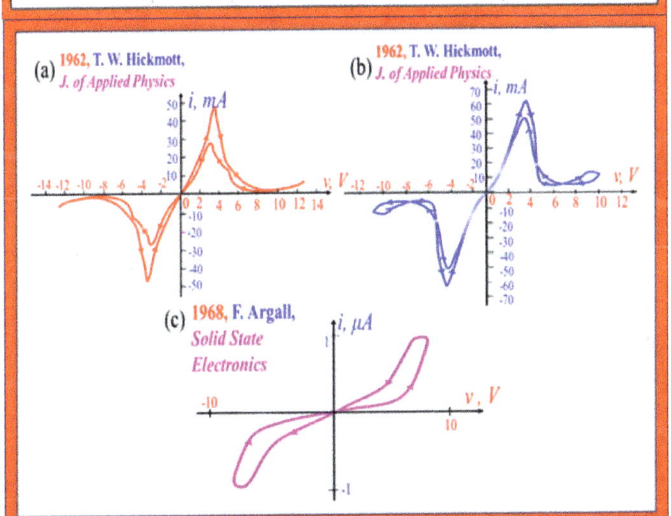
(a) 1962, T. W. Hickmott,
J. of Applied Physics
i, mA
v, V
(b) 1962, T. W. Hickmott,
J. of Applied Physics
i, mA
v, V
(c) 1968, F. Argall,
Solid State
Electronics
i, μA
v, V

3-33

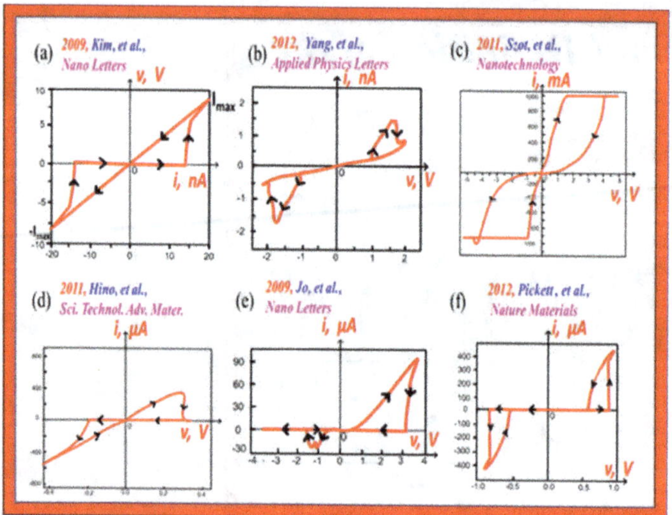
(a) 2009, Kim, et al., Nano Letters
v, V
i, nA
(b) 2012, Yang, et al., Applied Physics Letters
i, nA
v, V
(c) 2011, Szot, et al., Nanotechnology
i, mA
v, V
(d) 2011, Hino, et al., Sci. Technol. Adv. Mater.
i, μA
v, V
(e) 2009, Jo, et al., Nano Letters
i, μA
v, V
(f) 2012, Pickett, et al., Nature Materials
i, μA
v, V

3-34

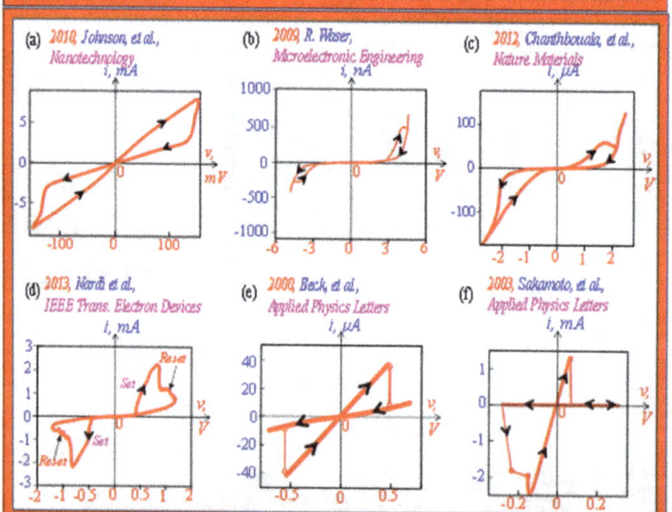
(a) 2010, Johnson, et al., Nanotechnology
i, mA
v, mV
(b) 2009, R. Waser, Microelectronic Engineering
i, nA
v, V
(c) 2012, Chanthbouala, et al., Nature Materials
i, μA
v, V
(d) 2013, Nardi et al., IEEE Trans. Electron Devices
i, mA
Reset
Set
v, V
(e) 2000, Beck, et al., Applied Physics Letters
i, μA
v, V
(f) 2003, Sakamoto, et al., Applied Physics Letters
i, mA
v, V

3-35

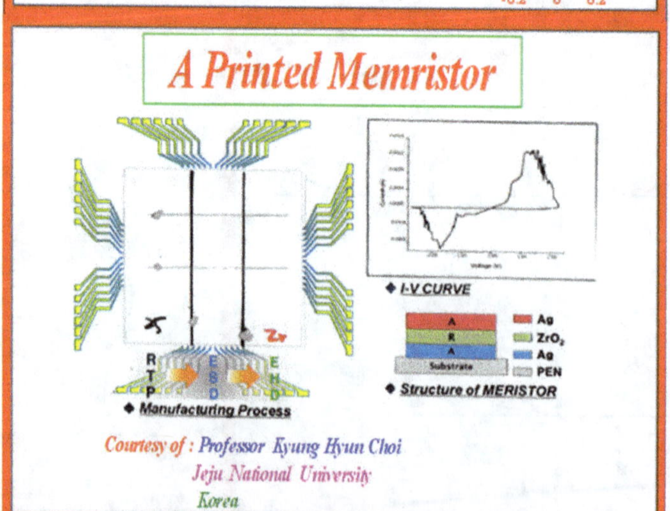
A Printed Memristor
R T P
E S D
E H D
◆ Manufacturing Process
◆ I-V CURVE
Ag
ZrO₂
Ag
PEN
Substrate
◆ Structure of MERISTOR
Courtesy of : Professor Kyung Hyun Choi
Jeju National University
Korea

3-36

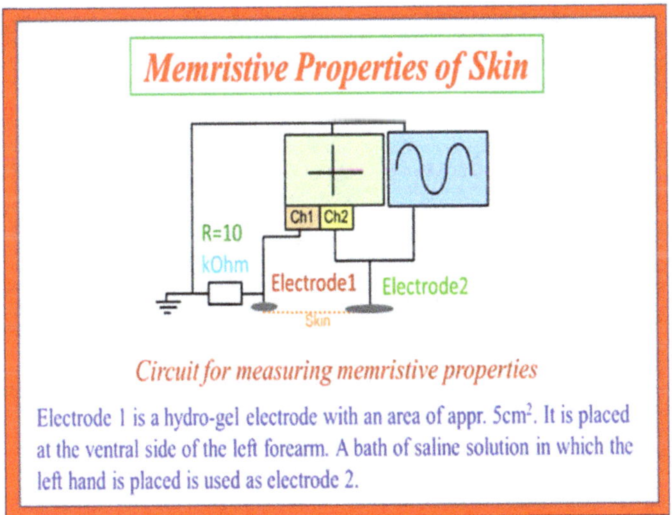
Memristive Properties of Skin
Ch1
Ch2
R=10
kOhm
Electrode1
Electrode2
Skin
Circuit for measuring memristive properties
Electrode 1 is a hydro-gel electrode with an area of appr. 5cm². It is placed at the ventral side of the left forearm. A bath of saline solution in which the left hand is placed is used as electrode 2.

3-37

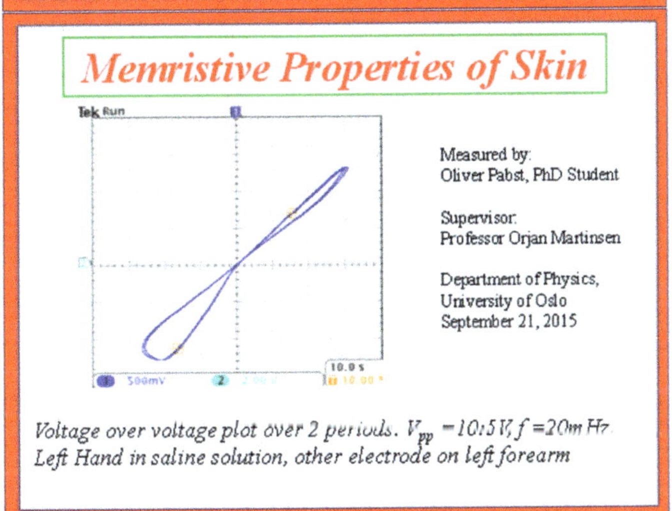
Memristive Properties of Skin
Tek Run
10.0 s
500mV
Measured by:
Oliver Pabst, PhD Student
Supervisor:
Professor Orjan Martinsen
Department of Physics,
University of Oslo
September 21, 2015
Voltage over voltage plot over 2 periods. Vpp =10.5V, f =20m Hz
Left Hand in saline solution, other electrode on left forearm

3-38

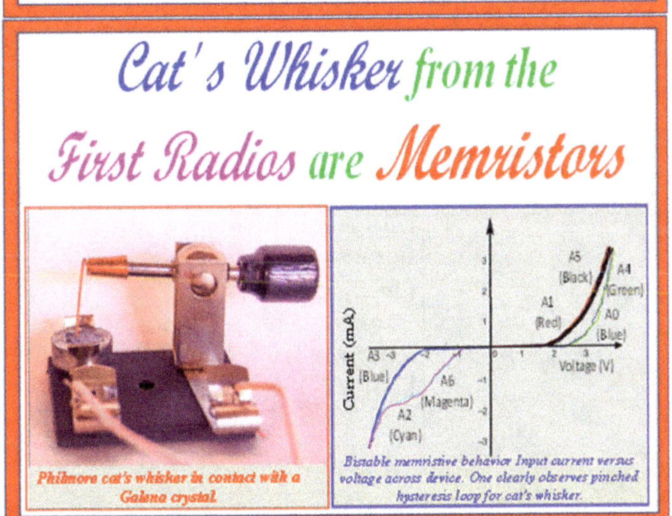
Cat's Whisker from the
First Radios are Memristors
Philmore cat's whisker in contact with a Galena crystal
Current (mA)
Voltage (V)
A5 (Black)
A4 (Green)
A1 (Red)
A0 (Blue)
A3 (Blue)
A6 (Magenta)
A2 (Cyan)
Bistable memristive behavior Input current versus voltage across device. One clearly observes pinched hysteresis loop for cat's whisker.

3-39

First Hint of Pinched Hysteresis Loop:

A Device Dubbed

LETTER 8 MEMORY

was published in 1971

POLARIZED (LETETR '8') MEMORY IN CdSe POINT CONTACT DIODES

M. Kikuchi, M. Saito, H. Okushi
Electrotechnical Laboratory, Mukodai, Tanashi, Tokyo, Japan.
and
A Matsuda
Nippon Columbia Co. Ltd., Kawasaki, Kanagawa, Japan.
(Received 10 March 1971 by T. Muto)

Solid State Communications
vol. 9, p. 705-707, 1971.

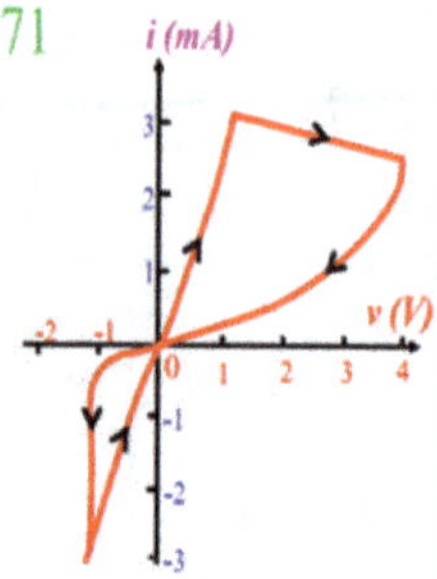

3-40

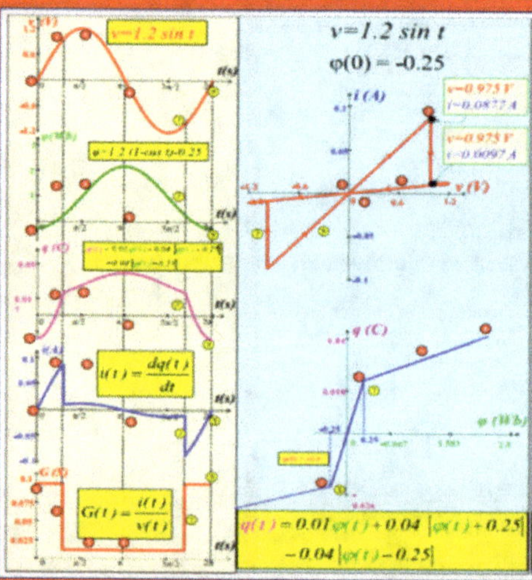

For enlargement of fine details, see Fig. 28, page 25 of Reference [10] in slide **1-195.**

3-41

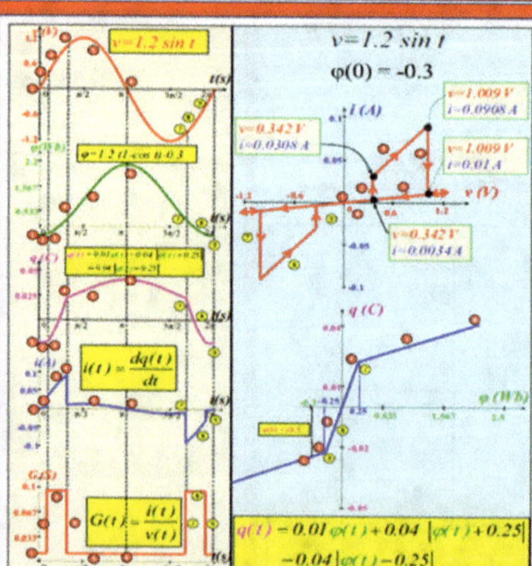

For enlargement of fine details, see Fig. 27, page 24 of Reference [10] in slide **1-195.**

3-42

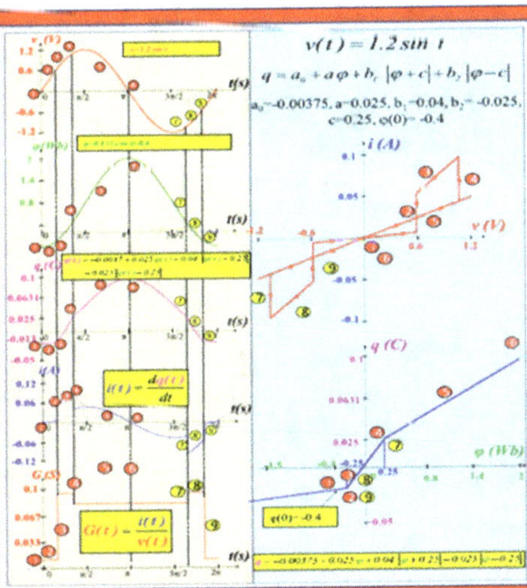

For enlargement of fine details, see Fig. 29, page 26 of Reference **[10]** in slide **1-195**.

3-43

Odd-Symmetric Hysteresis Loop Theorem

The Pinched hysteresis loop of every ideal memristor, and its ideal generic memristor siblings, derived, or measured, driven from a Sinusoidal voltage source v(t), or current source i(t) is Symmetric with Respect to the origin.

3-44

Closure Theorem

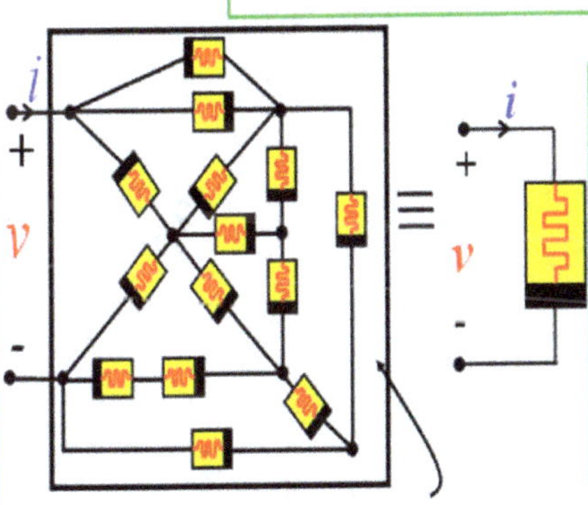

A one-port made of arbitrary interconnection of memristors is equivalent to a memristor.

3-45

Some Memristor Circuits have Hamiltonians ! 3-46

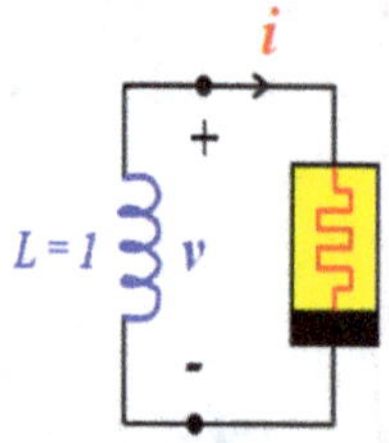

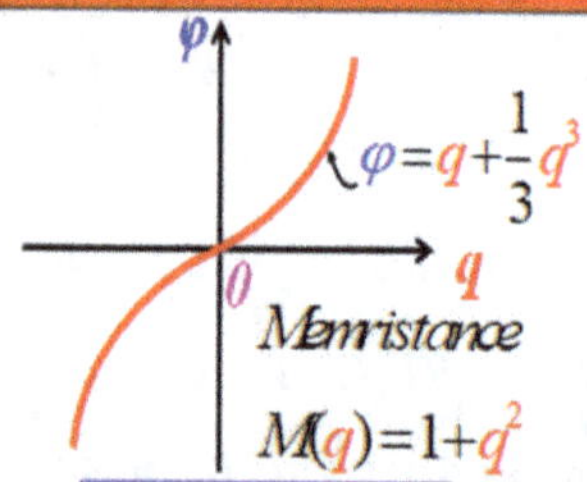

3-47

State Equations

$$\frac{dq}{dt} = i$$

$$\frac{di}{dt} = -(1+q^2)i$$

Hamiltonian

$$H(q,i) = \underbrace{(q + \frac{1}{3}q^3)}_{\text{Pseudo Potential Energy } H_q(q)} + \underbrace{i}_{\text{Pseudo Kinetic Energy } H_i(i)}$$

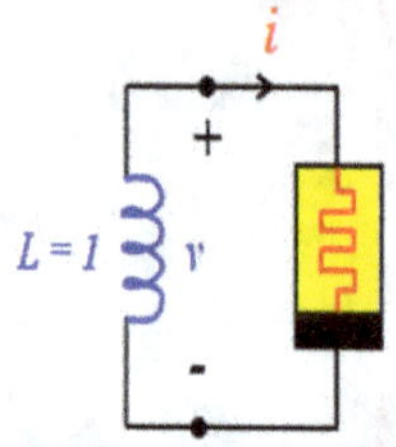

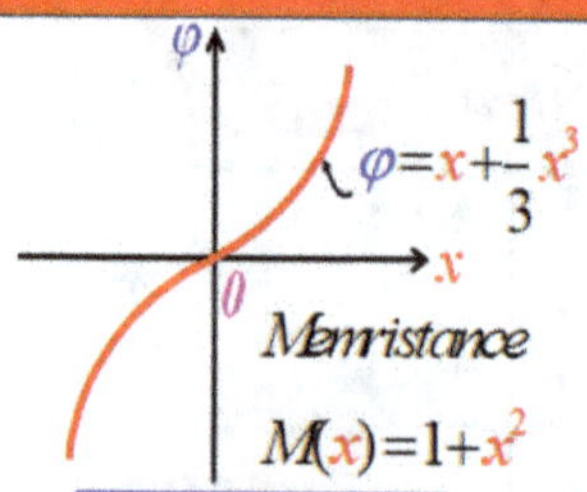

3-48

State Equations

$$\frac{dx}{dt} = i$$

$$\frac{di}{dt} = -(1+x^2)i$$

Hamiltonian

$$H(x,i) = \underbrace{(x + \frac{1}{3}x^3)}_{\text{Pseudo Potential Energy } H_x(x)} + \underbrace{i}_{\text{Pseudo Kinetic Energy } H_i(i)}$$

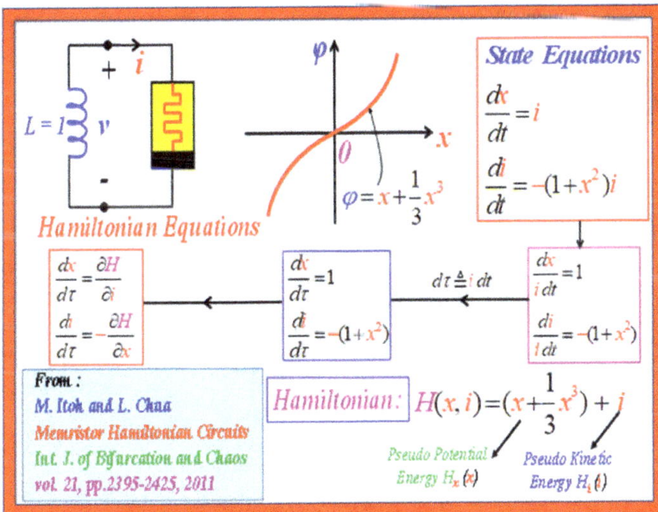

3-49

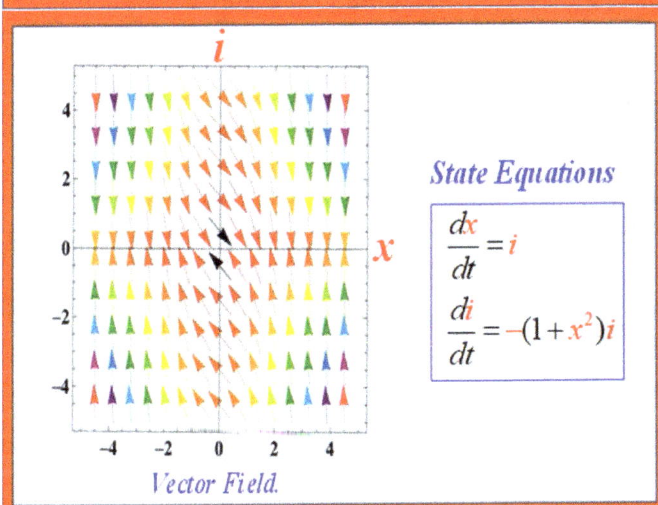

3-50

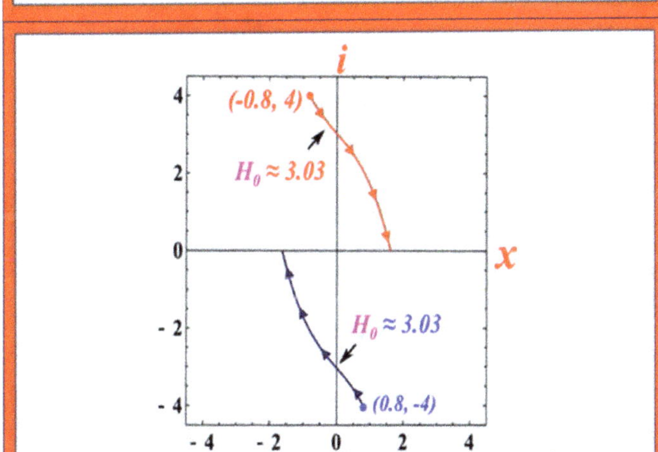

Two typical trajectories which tend to the x-axis along the contour $H_0 \approx 3.03$.

3-51

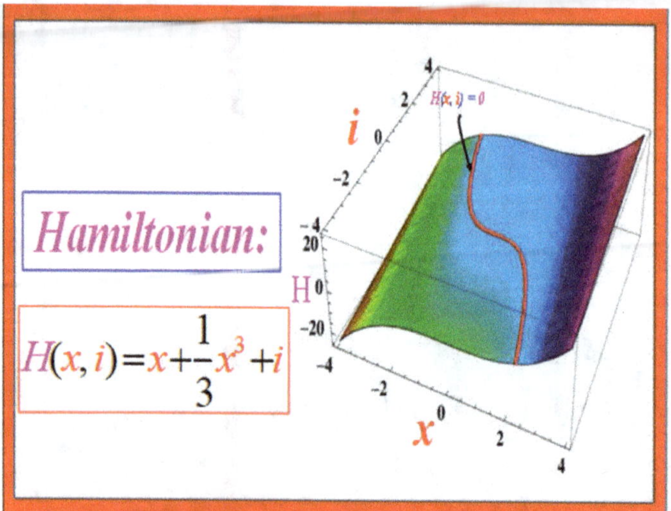
Hamiltonian:
H(x, i) = x + 1/3 x^3 + i
H(x, i) = 0
i
H
x

3-52

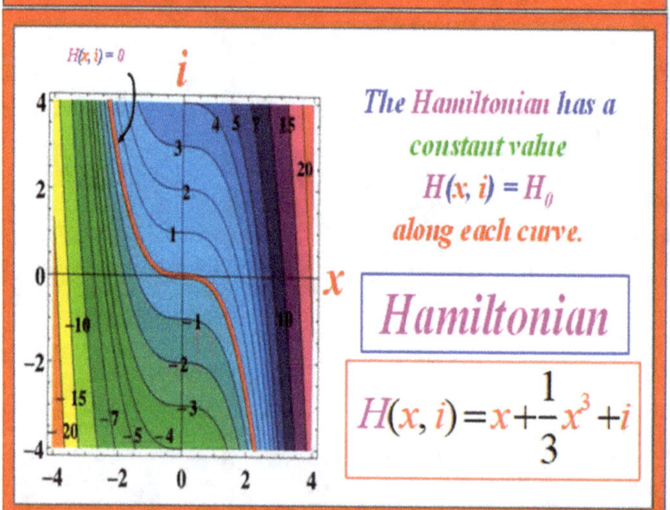
H(x, i) = 0
i
x
The Hamiltonian has a
constant value
H(x, i) = H0
along each curve.
Hamiltonian
H(x, i) = x + 1/3 x^3 + i

3-53

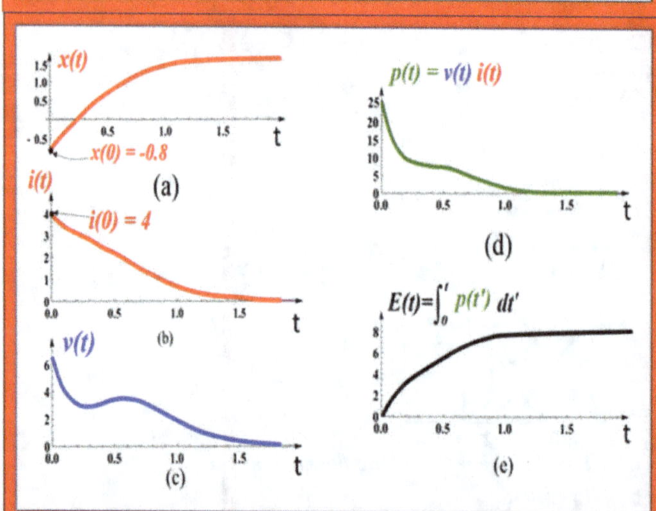
x(t)
x(0) = -0.8
(a)
i(t)
i(0) = 4
(b)
v(t)
(c)
p(t) = v(t) i(t)
(d)
E(t)=∫0t p(t') dt'
(e)
t

3-54

Pseudo Potential Energy

$$H_x = x + \frac{1}{3}x^3$$

Pseudo Kinetic Energy

$$H_i = i$$

Hamiltonian 3-55

$$H = \underbrace{H_x}_{\text{Pseudo Potential Energy}} + \underbrace{H_i}_{\text{Pseudo Kinetic Energy}}$$

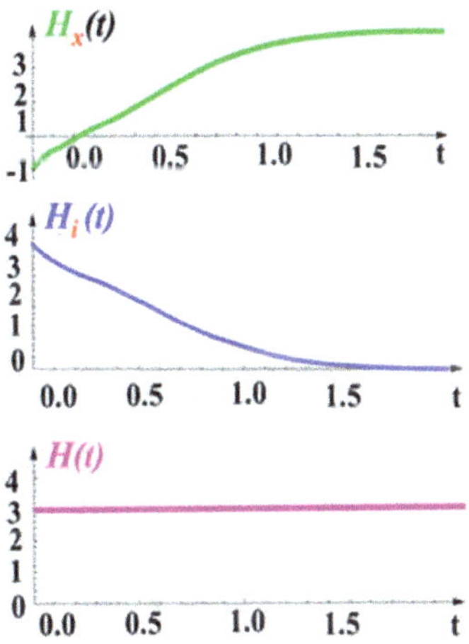

Shocking Revelation !

Not all

Hamiltonian Systems

are 3-56

Conservative

$ 1000 Prize

To the first person who gives a correct physical interpretation of the Hamiltonian in the Itoh-Chua memristor – inductor circuit and explains what physical entity is being conserved ? 3-57

Genealogy of Memristors

Extended Memristor

↓

Generic Memristor

↓

Ideal Generic Memristor

↓

Ideal Memristor

3-58

Constitutive Relation of Ideal Memristor

3-59

Current-controlled Ideal Memristor

$$\varphi = \hat{\varphi}(q)$$

Equivalent Representation :

$$v = R(q)\, i$$

$$\frac{dq}{dt} = i$$

Voltage-controlled Ideal Memristor

$$q = \hat{q}(\varphi)$$

Equivalent Representation :

$$i = G(\varphi)\, v$$

$$\frac{d\varphi}{dt} = v$$

Constitutive Relation of Ideal Memristor

3-60

Current-controlled Ideal Memristor

$$\varphi = \hat{\varphi}(x)$$

Equivalent Representation :

$$v = R(x)\, i$$

$$\frac{dx}{dt} = i$$

Voltage-controlled Ideal Memristor

$$q = \hat{q}(x)$$

Equivalent Representation :

$$i = G(x)\, v$$

$$\frac{dx}{dt} = v$$

Constitutive Relation of Ideal Generic Memristor 3-61

Current-controlled Ideal Generic Memristor

$$v = R(x)\, i$$

$$\frac{dx}{dt} = \hat{f}(x)\, i$$

where

$$F(x) = \int \left[\frac{1}{\hat{f}(x)} \right] dx$$

is a one-to-one function

Voltage-controlled Ideal Generic Memristor

$$i = G(x)\, v$$

$$\frac{dx}{dt} = \hat{g}(x)\, v$$

where

$$G(x) = \int \left[\frac{1}{\hat{g}(x)} \right] dx$$

is a one-to-one function

An Illustrative Example Creating An Ideal Generic Memristor Siblings from an Ideal Memristor 3-62

Ideal Memristor

$$\varphi = q + \frac{1}{3} q^3$$

State-Dependent Ohm's Law :

$$v = (1 + q^2)\, i$$

State Equation : $\frac{dq}{dt} = i$

Ideal Generic Memristor Sibling

$$\varphi = x^{1/3} + \frac{1}{3} x$$

State-Dependent Ohm's Law : $v = (1 + x^{1/3})\, i$

State Equation :

$$\frac{dx}{dt} = 3(x^{2/3})\, i$$

Recovering Ideal Memristor from its Sibling ? 3-63

Given:

State-Dependent Ohm's Law: $v = R(x)\, i$

State Equation: $\frac{dx}{dt} = f(x)\, i$; where $f(x) \geq 0$

Step: 1

Calculate $q = \int \left[\frac{1}{f(x)} \right] dx \triangleq h(x)$

Step: 2

Calculate *inverse function*:

$$x = h^{-1}(q)$$

Step: 3

Substitute $h^{-1}(q)$ for x in $R(x)$:

$$R(x)\Big|_{x = h^{-1}(q)} \triangleq \hat{R}(q)$$

Step: 4

Ideal Memristor is describe by

$$v = \hat{R}(q)\, i \qquad \frac{dq}{dt} = i$$

Example

Given:

State-Dependent Ohm's Law:

$$v = \underbrace{(1 + x^{2/3})}_{R(x)}\, i$$

State Equation:

$$\frac{dx}{dt} = \underbrace{(3x^{2/3})}_{f(x)}\, i$$

Step: 1

$$q = \int \left[\frac{1}{3x^{2/3}} \right] dx = x^{1/3} \triangleq h(x)$$

Step: 2

$$x = h^{-1}(q) = q^3$$

Step: 3

$$R(x)\Big|_{x = q^3} = (1 + x^{2/3})\Big|_{x = q^3} = 1 + q^2 \triangleq \hat{R}(q)$$

Step: 4 *Ideal Memristor*

$$v = (1 + q^2)\, i \qquad \frac{dq}{dt} = i$$

Memristor State Variable Transform Theorem

Two Memristors whose state variables are related by a *one-to-one piecewise-differentiable transformations* have *identical pinched hysteresis loops*

3-64

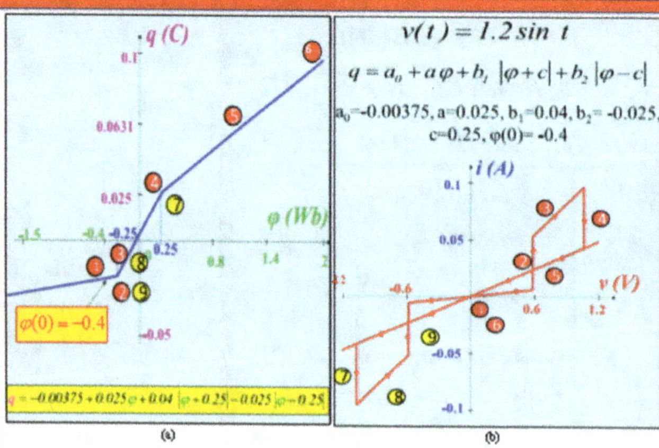

For enlargement of fine details, see Fig. 29, page 26 of Reference **[10]** in slide **1-195**.

3-65

A 1:1 Piecewise-Differentiable Function and Its Inverse Function

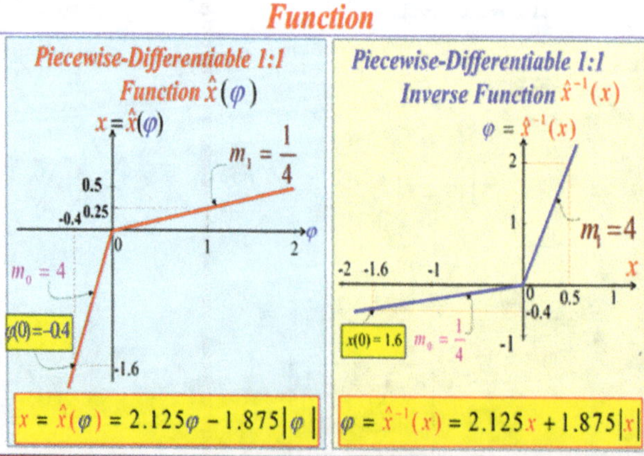

3-66

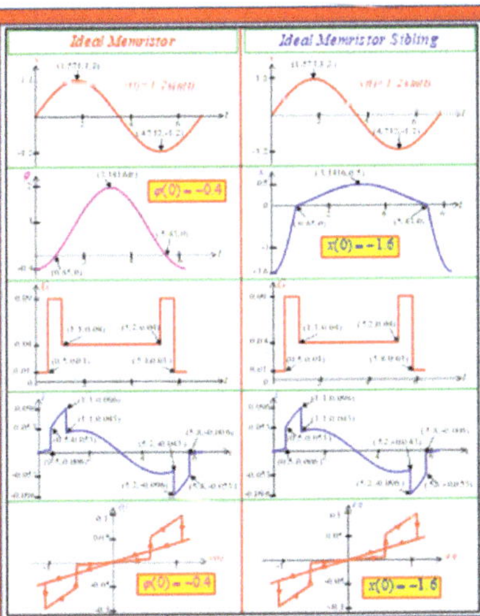

For enlargement of fine details, see Fig. 15, page 337 of Reference **[11]** in slide **1-195**. 3-67

HP TiO_2 Memristor is an *Ideal Generic Memristor* 3-68

Consider following *Ideal Memristor* :

State-Dependent Ohm's Law: $v = M(q)\, i$ (1)

where, $M(q) \triangleq R_{OFF}\left[1 - \frac{\mu_v R_{ON}}{D^2}\right] q$

State Equation: $\frac{dq}{dt} = i$ (2)

Let us define another state variables w via the following *one-to-one* transformation:

$w = Kq$, $q = 1/K\; w$ (3)

where, $K \triangleq \frac{\mu_v R_{ON}}{D}$ (4)

Then, $v = \hat{R}(w)\, i$ (5) where, $\frac{dw}{dt} = K\frac{dq}{dt} = Ki$

$\hat{R}(w) \triangleq R_{OFF}\left[1 - \frac{\mu_v R_{ON}}{D^2}\right]\left[\frac{D}{\mu_v R_{ON}}\right] w$ (6)

$$\frac{dw}{dt} = \underbrace{\left(\frac{\mu_v R_{ON}}{D}\right)}_{\hat{f}(w)} i \quad (7)$$

Equations (5) and (7) defines an Ideal Generic Memristor.

Every

Ideal Memristor

spawns an

Infinite family

of 3-69

Equivalent

Generic Memristor

Siblings

Zoo of *Ideal Generic Memristors* 3-70

Constitutive Relation of Generic Memristor 3-71

Current-controlled Generic Memristor	*Voltage-controlled Generic Memristor*
$v = R(\mathbf{x})i$ $\frac{d\mathbf{x}}{dt} = \mathbf{f}(\mathbf{x}, i)$	$i = G(\mathbf{x})v$ $\frac{d\mathbf{x}}{dt} = \mathbf{g}(\mathbf{x}, v)$

Example : Generic Memristor 3-72

State-Dependent Ohm's Law

$$v = \underbrace{\left(\frac{1}{4}x\right)}_{R(x)} i$$

Memristance ⇨ $R(x)$

State Equation

$$\frac{dx}{dt} = \underbrace{x - i^2}_{f(x,i)}$$

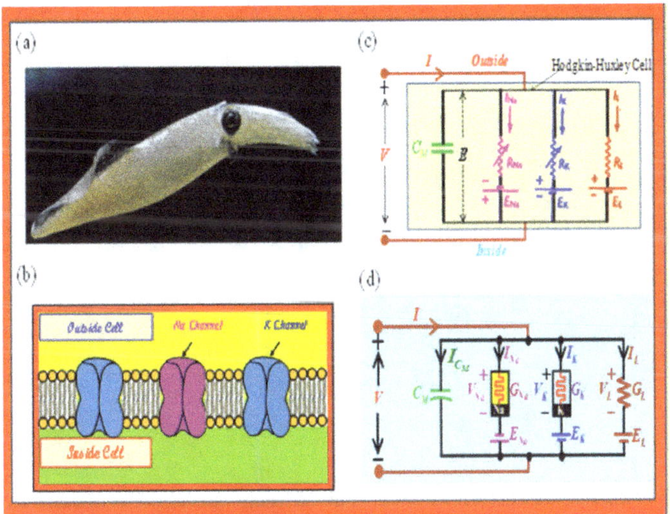
(a)
(b)
(c)
(d)
Outside Cell
Na Channel
K Channel
Inside Cell
Outside
Hodgkin-Huxley Cell
Inside

3-73

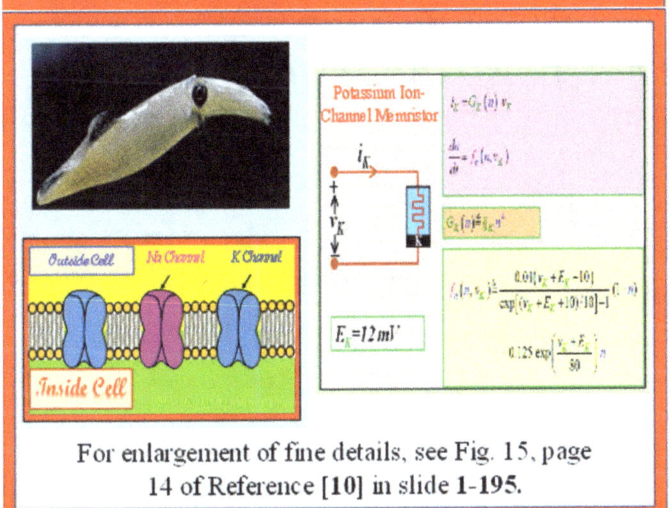
Potassium Ion-Channel Memristor
Outside Cell
Na Channel
K Channel
Inside Cell
For enlargement of fine details, see Fig. 15, page 14 of Reference [10] in slide 1-195.

3-74

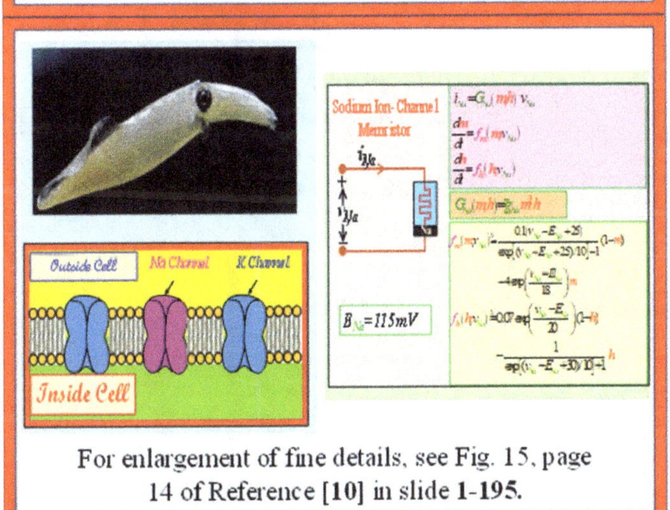
Sodium Ion-Channel Memristor
Outside Cell
Na Channel
K Channel
Inside Cell
For enlargement of fine details, see Fig. 15, page 14 of Reference [10] in slide 1-195.

3-75

Constitutive Relation of Extended Memristor 3-76

Current-controlled Extended Memristor

$$v = R(\mathbf{x}, i)\,i \qquad R(\mathbf{x}, 0) \neq \infty$$

$$\frac{d\mathbf{x}}{dt} = \mathbf{f}(\mathbf{x}, i)$$

Voltage-controlled Extended Memristor

$$i = G(\mathbf{x}, v)\,v \qquad G(\mathbf{x}, 0) \neq \infty$$

$$\frac{d\mathbf{x}}{dt} = \mathbf{g}(\mathbf{x}, v)$$

Example : Extended Memristor 3-77

Ohm's Law

$$v = \underbrace{\left(0.01 x^2 i^2\right)}_{R(x,\, i)} i$$

State Equation

$$\frac{dx}{dt} = \underbrace{-x^3 - 2x^2 + (3 + i^2)x}_{f(x,\, i)}$$

Example of an Extended Memristor 3-78

Relatively Low Frequencies

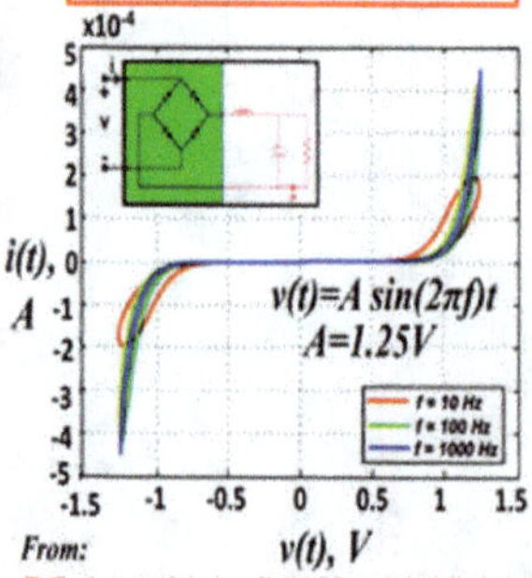

Very High Frequencies

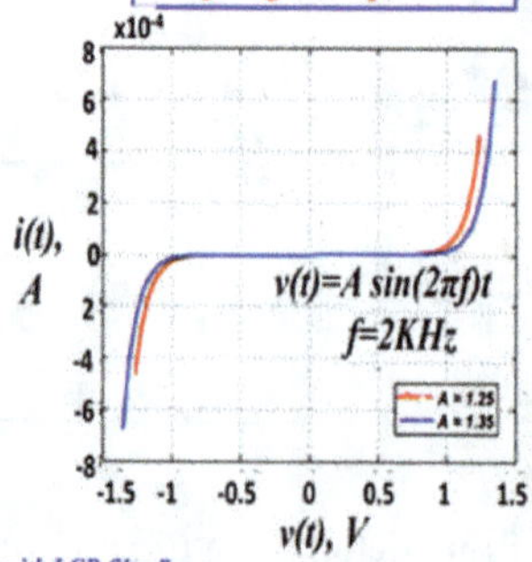

From:
F. Corinto and A. Ascoli, "A Memristive diode bridge with LCR filter", Electronics Letters, vol. 48, no.14, pp.824-825, 2012

Genealogy of Memristors

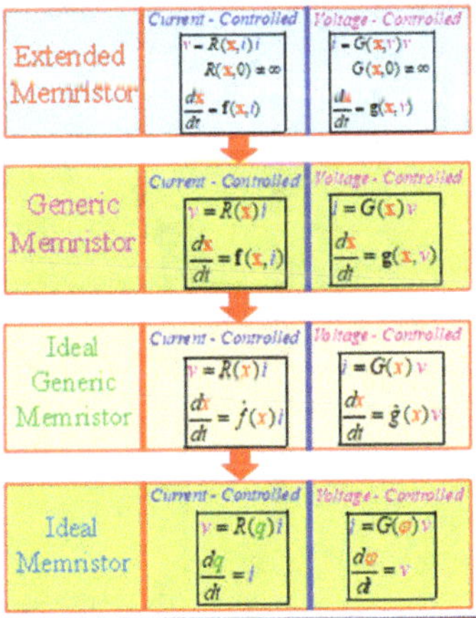

For enlargement of fine details, see Table 1 on page 323 of Reference [11] in slide **1-195**. 3-79

The Memristor Universe

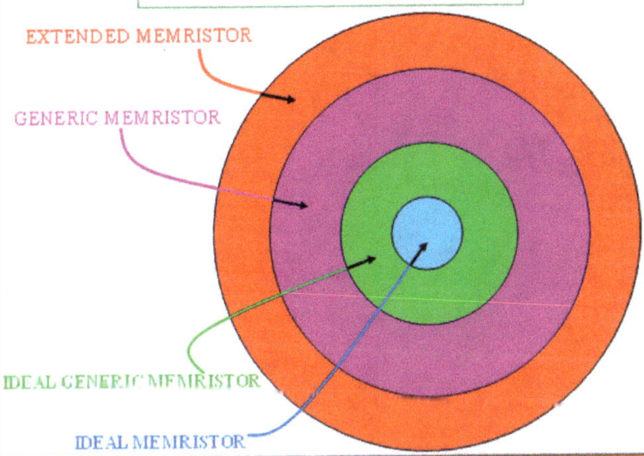

3-80

Why is the Condition

$$R(x,0) \neq \infty$$

necessary in the definition of the state and Current-Dependent Ohm's Law 3-81

$$v = R(x, i)\, i$$

?

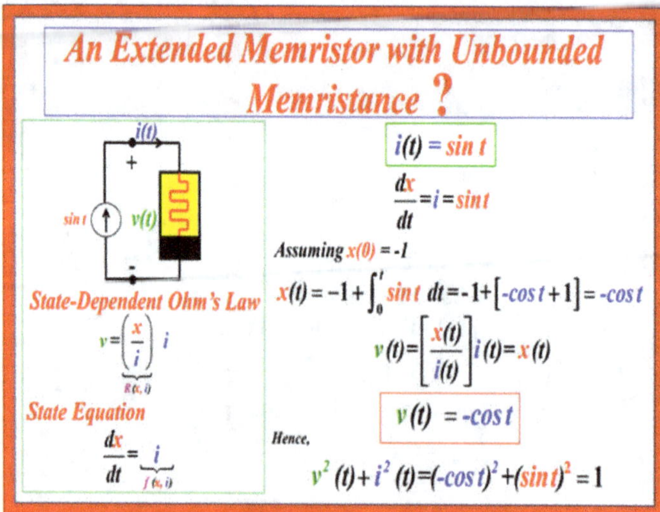

An Extended Memristor with Unbounded Memristance ?
i(t)
+
sin t
v(t)
-
State-Dependent Ohm's Law
$v = \underbrace{\left(\frac{x}{i}\right)}_{R(x,i)} i$
State Equation
$\frac{dx}{dt} = \underbrace{i}_{f(x,i)}$
$i(t) = \sin t$
$\frac{dx}{dt} = i = \sin t$
Assuming $x(0) = -1$
$x(t) = -1 + \int_0^t \sin t \, dt = -1 + [-\cos t + 1] = -\cos t$
$v(t) = \left[\frac{x(t)}{i(t)}\right] i(t) = x(t)$
$v(t) = -\cos t$
Hence,
$v^2(t) + i^2(t) = (-\cos t)^2 + (\sin t)^2 = 1$

3-82

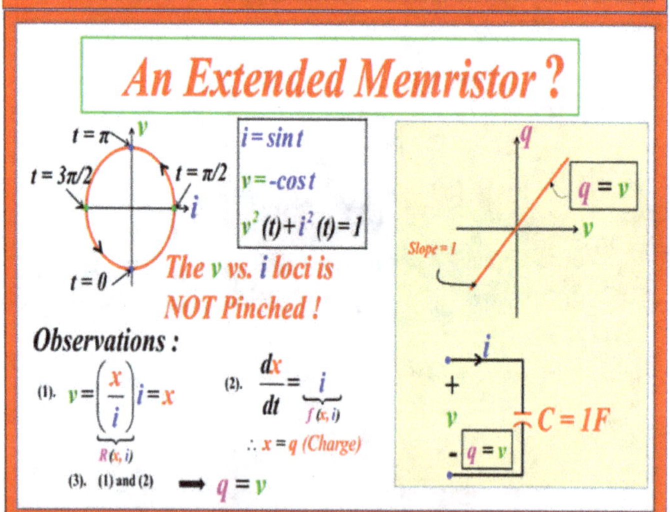

An Extended Memristor ?
$t = \pi$
v
$t = 3\pi/2$
$t = \pi/2$
i
$t = 0$
$i = \sin t$
$v = -\cos t$
$v^2(t) + i^2(t) = 1$
The v vs. i loci is NOT Pinched !
q
q = v
v
Slope = 1
i
+
v
C = 1F
-
q = v
Observations :
(1). $v = \underbrace{\left(\frac{x}{i}\right)}_{R(x,i)} i = x$
(2). $\frac{dx}{dt} = \underbrace{i}_{f(x,i)}$
$\therefore x = q$ (Charge)
(3). (1) and (2) ➡ $q = v$

3-83

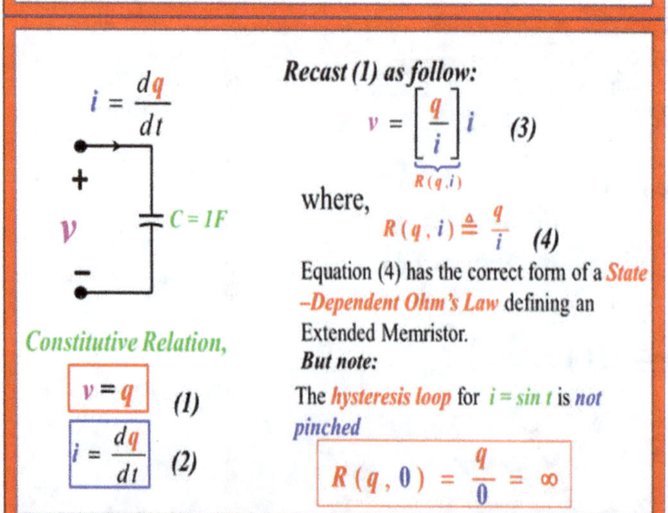

$i = \frac{dq}{dt}$
+
v
C = 1F
-
Constitutive Relation,
$v = q$ (1)
$i = \frac{dq}{dt}$ (2)
Recast (1) as follow:
$v = \underbrace{\left[\frac{q}{i}\right]}_{R(q,i)} i$ (3)
where,
$R(q,i) \triangleq \frac{q}{i}$ (4)
Equation (4) has the correct form of a State-Dependent Ohm's Law defining an Extended Memristor.
But note:
The hysteresis loop for $i = \sin t$ is not pinched
$R(q,0) = \frac{q}{0} = \infty$

3-84

If It's Pinched

It's A 3-85

MEMRISTOR

If It's Not Pinched

It's Not A 3-86

MEMRISTOR

Fingerprints

Of 3-87

Memristors

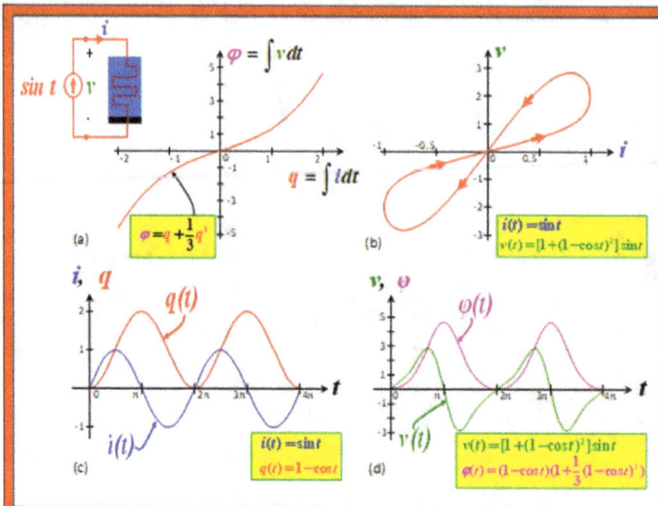

3-88

Ideal Memristor Symmetry Theorem

The pinched hysteresis loop associated with an Ideal Memristor, or any of its Ideal generic memristor siblings, under sinusoidal excitation, is always odd symmetric.

3-89

How does a Memristor behave at very high frequencies ?

3-90

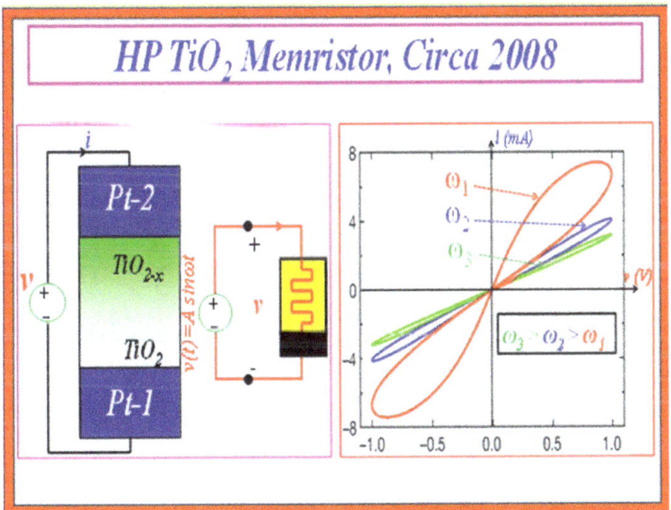
HP TiO2 Memristor, Circa 2008
Pt-2
TiO2-x
TiO2
Pt-1
v(t)=A sinωt
i (mA)
v (V)
ω3 > ω2 > ω1

3-91

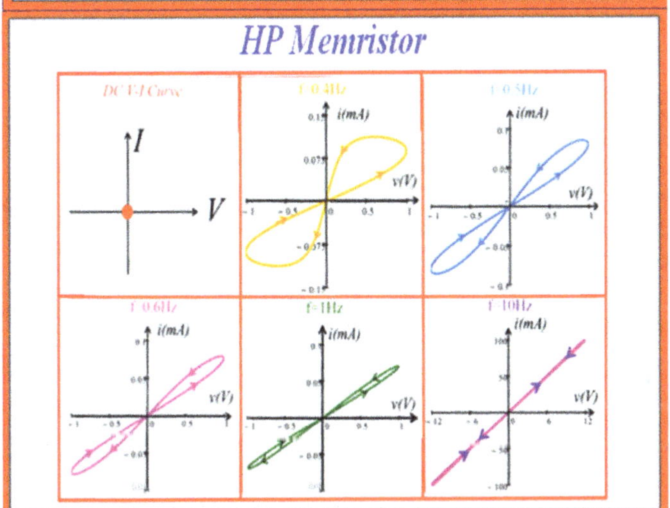
HP Memristor
DC V-I Curve
I
V
i(mA)
v(V)

3-92

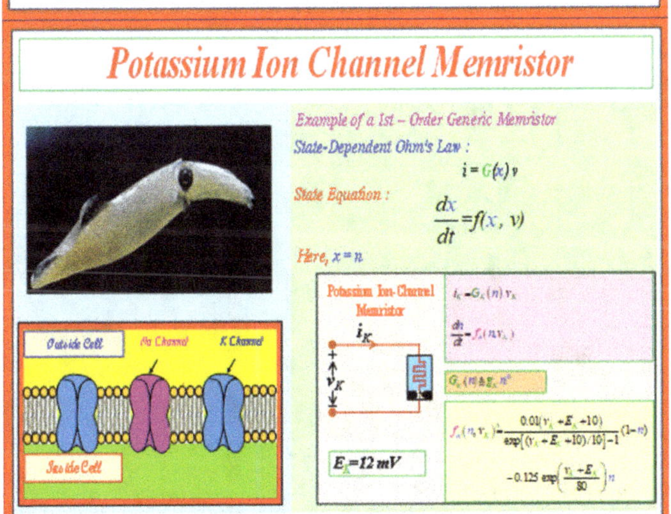
Potassium Ion Channel Memristor
Example of a 1st – Order Generic Memristor
State-Dependent Ohm's Law :
i = G(x) v
State Equation :
dx/dt = f(x, v)
Here, x = n
Potassium Ion-Channel Memristor
E_K=12 mV
Outside Cell
Na Channel
K Channel
Inside Cell

3-93

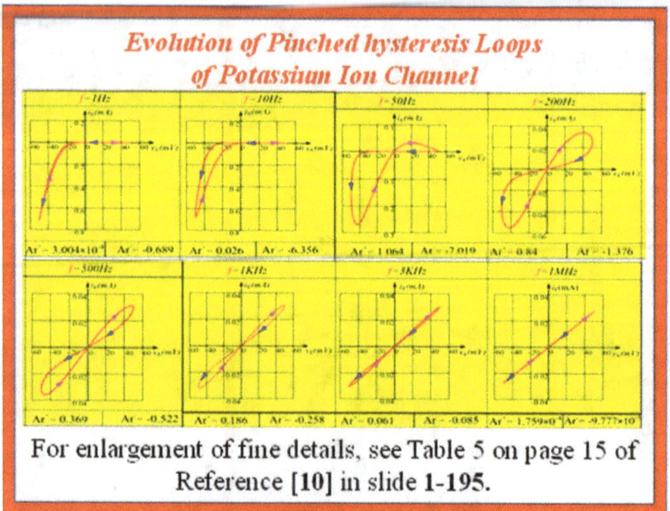
Evolution of Pinched hysteresis Loops
of Potassium Ion Channel
For enlargement of fine details, see Table 5 on page 15 of
Reference [10] in slide 1-195.

3-94

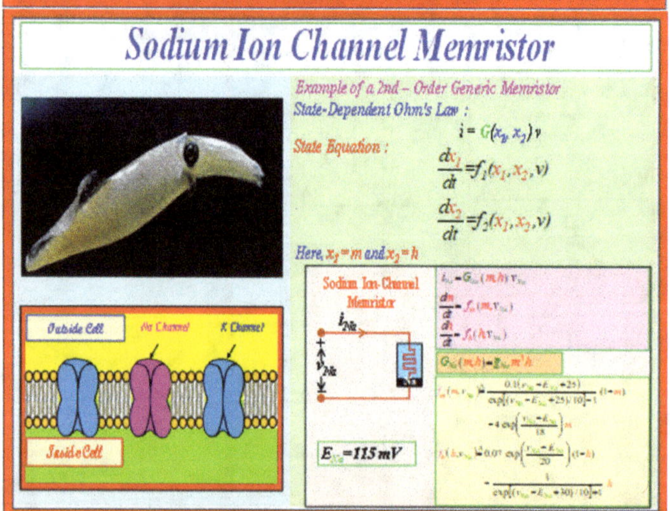
Sodium Ion Channel Memristor
Example of a 2nd – Order Generic Memristor
State-Dependent Ohm's Law :
$i = G(x_1, x_2)v$
State Equation :
$\frac{dx_1}{dt} = f_1(x_1, x_2, v)$
$\frac{dx_2}{dt} = f_2(x_1, x_2, v)$
Here, $x_1 = m$ and $x_2 = h$
Outside Cell
Na Channel
K Channel
Inside Cell
Sodium Ion Channel Memristor
$E_{Na} = 115\,mV$

3-95

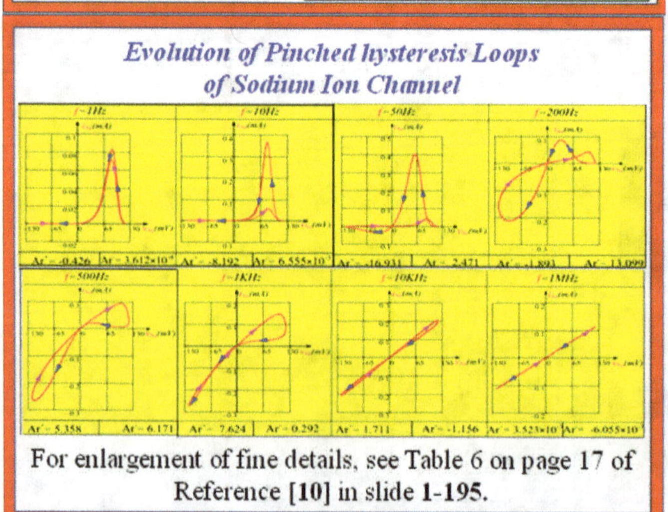
Evolution of Pinched hysteresis Loops
of Sodium Ion Channel
For enlargement of fine details, see Table 6 on page 17 of
Reference [10] in slide 1-195.

3-96

Fingerprints of Generic Memristors 3-97

Beyond some frequency ω_0, the lobe area of the pinched hysteresis loop of any

Generic Memristor

shrinks continuously as the frequency ω increases, such that the pinched hysteresis loop tends to a straight line with slope m_∞ (which depends on the amplitude and initial state) at sufficiently high frequencies.

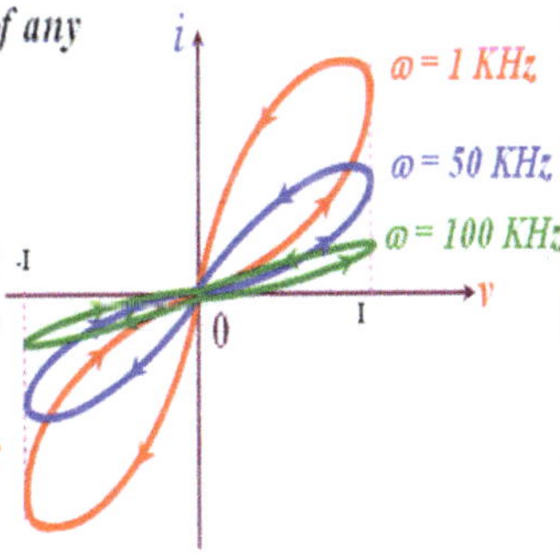

Fingerprints of Extended Memristors 3-98

Beyond some frequency ω_0, the lobe area of the pinched hysteresis loop of any

Extended Memristor

shrinks and deforms continuously as the frequency ω increases, such that the pinched hysteresis loop tends to a

Single-valued function

in the v-i plane.

Pinched Hysteresis Loop of An Extended Memristor 3-99

Example of a 1st – Order Extended Memristor

State-Dependent Ohm's Law : $v = R(x, i)\, i = (0.01\, x^2 i^2)\, i$

State Equation : $\frac{dx}{dt} = -x^3 - 2x^2 + (3 + i^2)\, x$

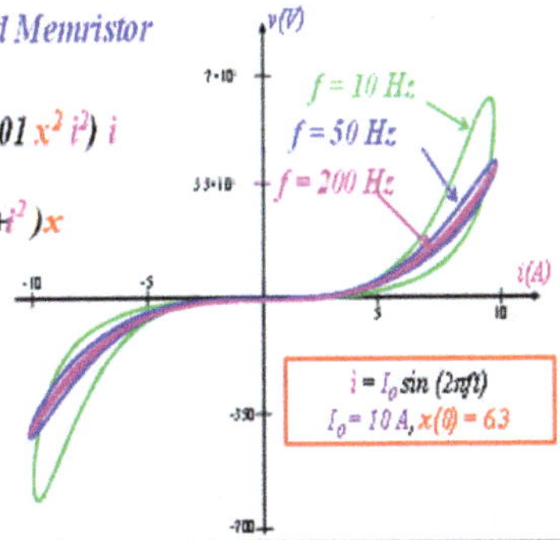

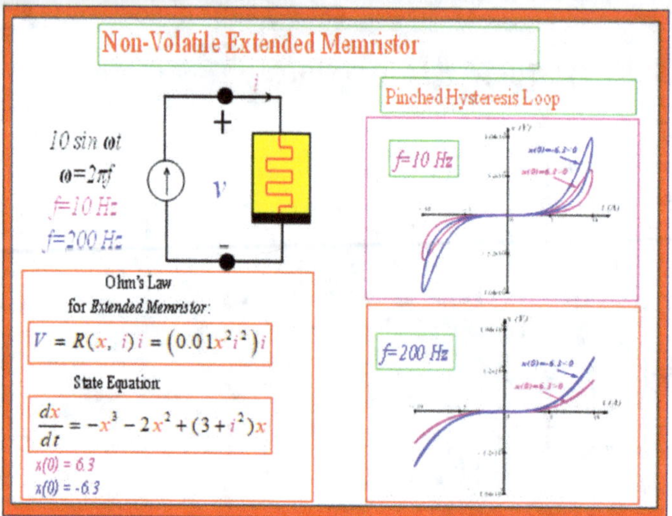

3-100

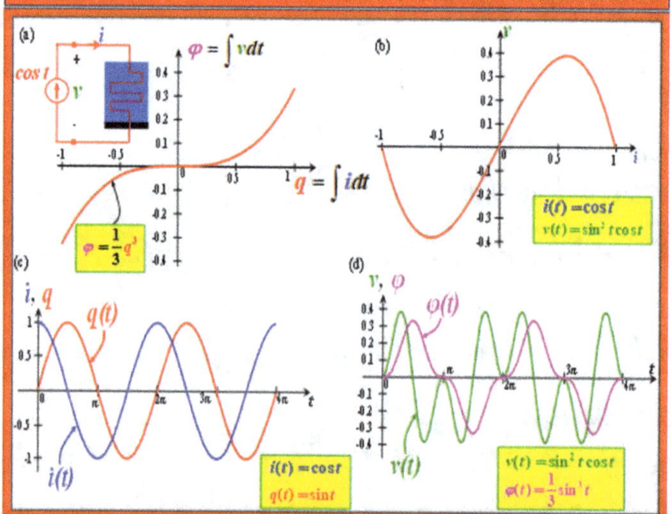

3-101

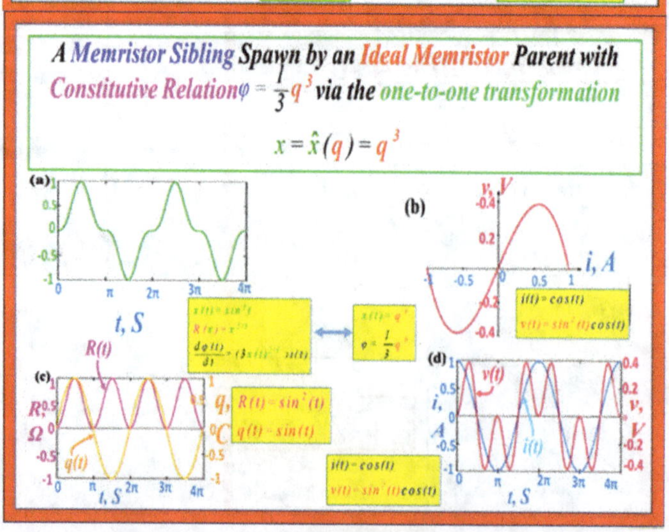

3-102

Pinched Hysteresis Loops

are

NOT

models

because they

can not predict

the response due to arbitrary

input waveforms 3-103

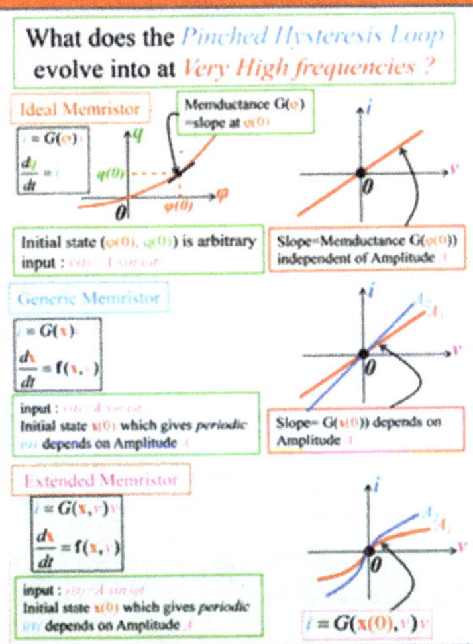

3-104

Characteristic Fingerprints of All Memristors

Above some critical frequency

$f = f^*$

the lobe area of the

pinched hysteresis loop

decreases monotonically,

until it shrinks to a straight line, for

a generic memristor or a single-valued curve for an

extended memristor. 3-105

Exclusive Property of Extended Memristors

Pinched hysteresis loop tends to a *single-valued* nonlinear function as $\omega \to \infty$ 3-106

Testing For Memristor Type

1. If *pinched hysteresis loop* is *not odd symmetric*,

   it is not an ideal memristor

2. If *pinched hysteresis loop* tends to a *straight line* whose slope *does not depend* on the *amplitude of the sinusoidal testing signal* as $\omega \to \infty$

   it is an ideal memristor 3-107

3. If *pinched hysteresis loop* tends to a *straight line* whose *slope depend* on the *amplitude of the sinusoidal testing signal* as $\omega \to \infty$

   it is a Generic memristor

4. If *pinched hysteresis loop* tends to a *single-valued nonlinear function* as $\omega \to \infty$

   it is an Extended Memristor

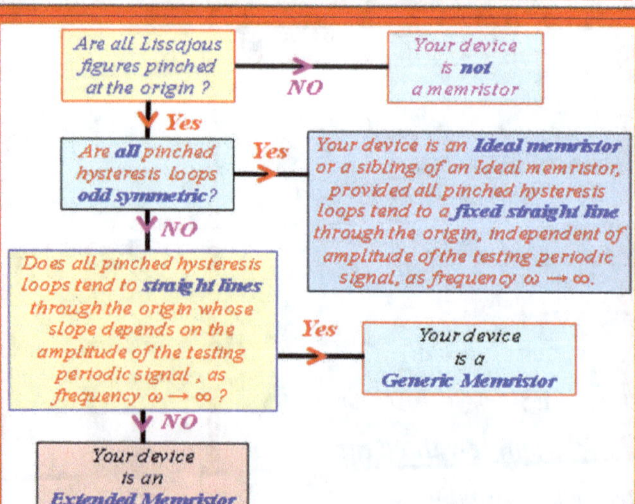

3-108

3-109

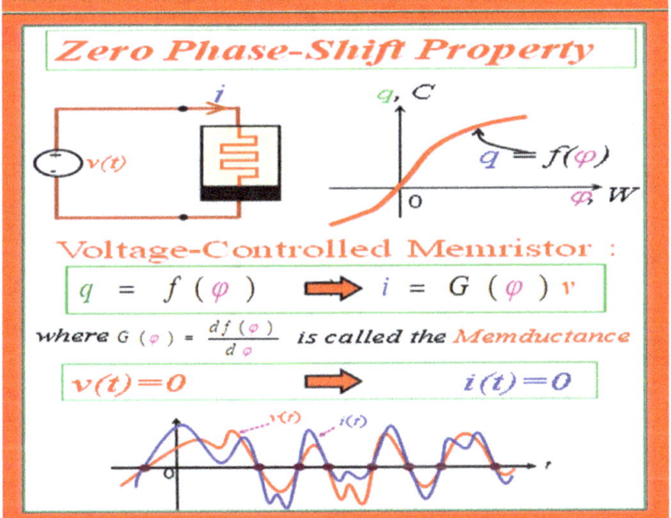

3-110

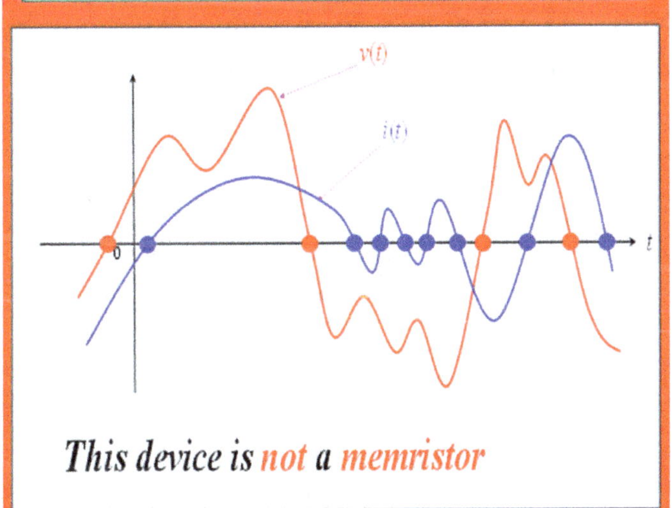

3-111

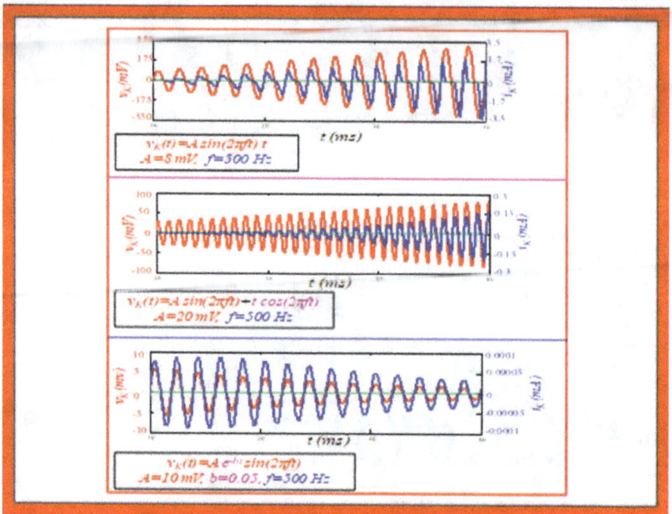

3-112

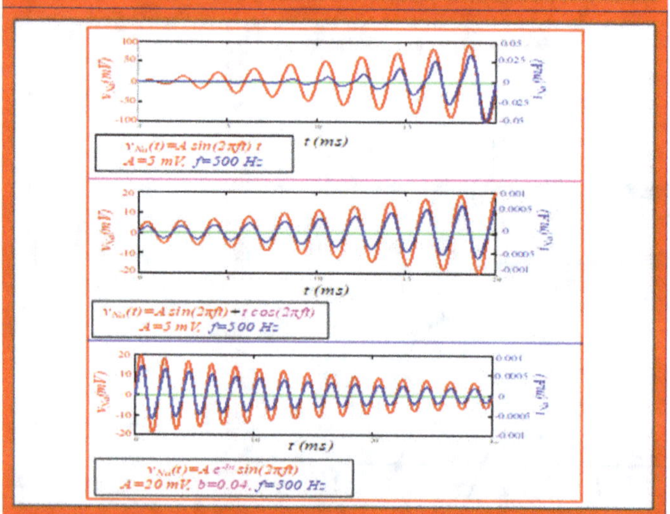

3-113

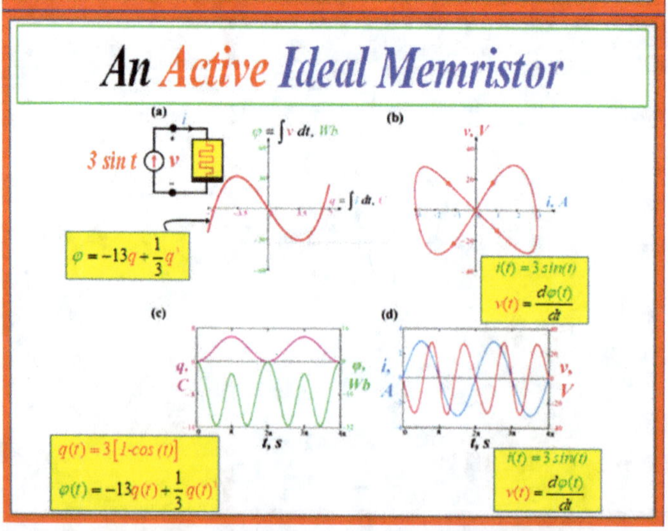

3-114

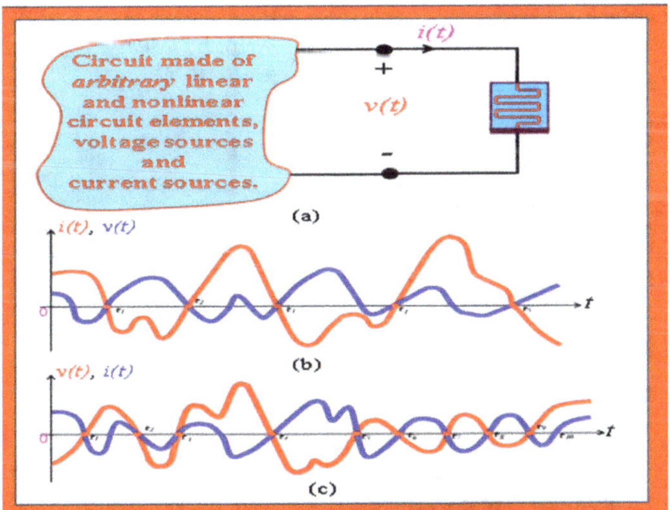

3-115

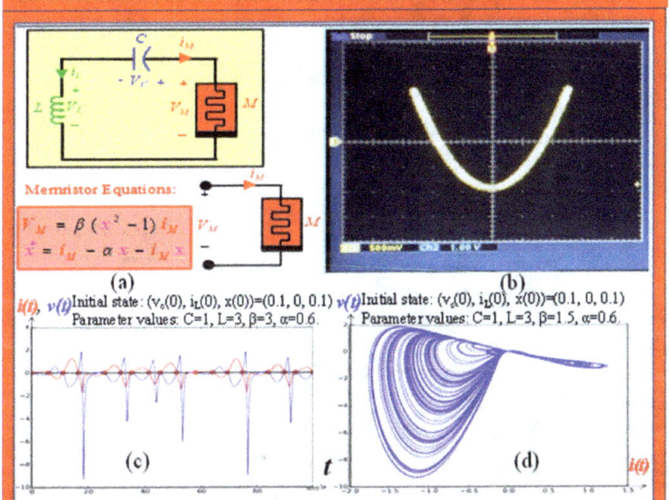

3-116

Ideal Memristor, Generic Memristors, and

Extended Memristors

are

memristors

because they all exhibits the same characteristic

Fingerprints

and

Signatures

3-117

The First Man-Made Memristor

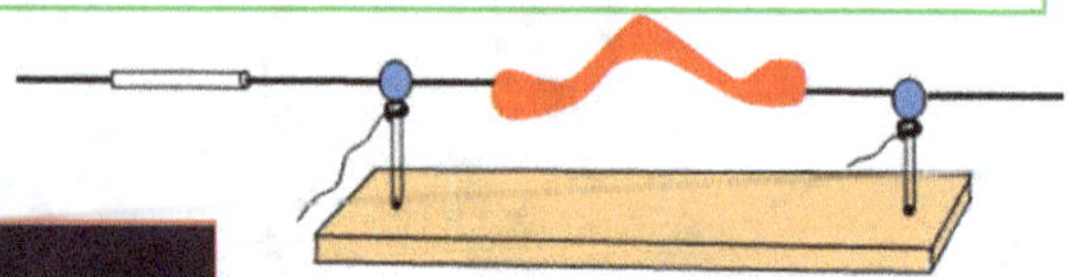

Sir Humphry Davy

ELEMENTS
of
CHEMICAL PHILOSOPHY
^^^^^^^^^^^^^^^^
BY
SIR HUMPHRY DAVY, LL D.
First Successful Demonstration of Continuous arc Memristor :
1801

3-118

3-119

Alessandro Volta

In honor of his invention **Volta** was made a *count* by Napoleon in **1801**.

Volta explains the principle of the "electrode column" to Napoleon

Voltaic Pile
(Invented by Volta
1800)

3-120

Setup of the carbon rod electrodes

3-121

Carbon arc discharge in operation

3-122

Davy's Carbon-Rod Electrodes Exhibits the Memristor Pinched-Hysteresis Loop Fingerprints

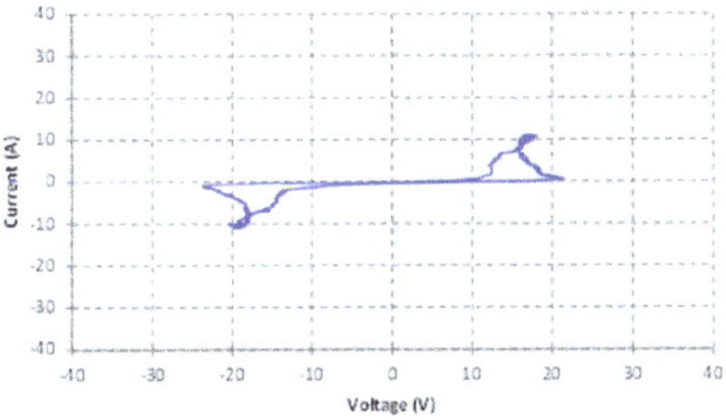

Input Testing Signal: 3 KHz Sinusoidal Voltage

3-123

Davy's Carbon-Rod Electrodes Exhibits the Memristor Zero Phase-shift signature 3-124

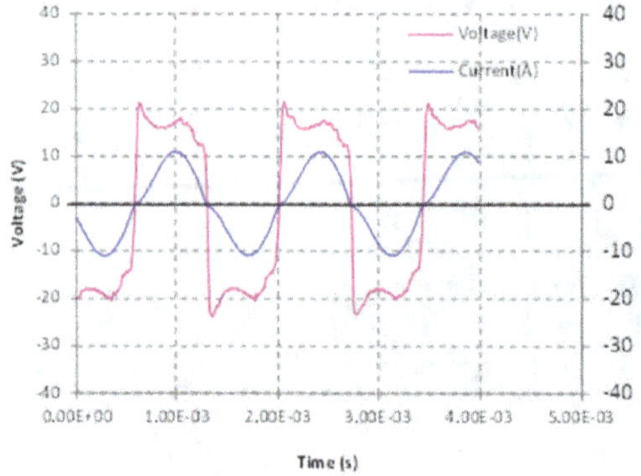

Videos 3-125

1

2

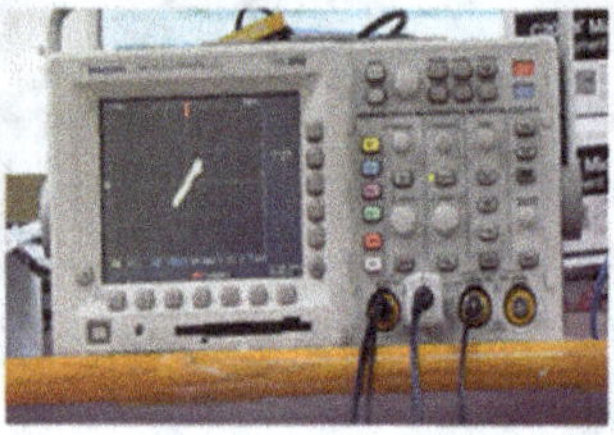

Bonaparte Napoleon

Sir Humphry Davy

3-126

In 1808, when France was at war with England, *Bonaparte Napoleon* has decided to award *Sir Humphry Davy*

The Prix Napoleon de Institut !

Faraday Invented the *Inductor* in 1831

Michael Faraday

3-127

10 Things You Didn't Know About Memristors

1. Every *ideal memristor* is a *non-volatile memory*.
2. Every *ideal memristor* is endowed with a *continuum* of *equilibrium states*.
3. An *ideal memristor* has *no transient* component in its response to any *periodic input signal*.
4. The *shape* of the *pinched hysteresis loop* of an *ideal memristor* can change drastically by changing only the *initial condition*.

3-128

10 Things You Didn't Know About Memristors

5. The DC V-I curve of an *ideal memristor* consists of only *one point*; namely, $(V, I) = (0, 0)$.
6. The *shape* of the *pinched hysteresis loop* of all *memristors* changes with the *amplitude*, or *frequency*, of the *periodic input signal*.
7. Every *ideal memristor* spawns an *infinite family* of *equivalent ideal generic memristor siblings*.

3-129

10 Things You Didn't Know About Memristors

8. All *memristor* siblings spawned from an *ideal memristor* parent are *indistinguishable* from their parent in the sense that they have *identical pinched hysteresis loops* when driven from the same *input signal* and *initial condition*.

9. The *hp TiO_2 memristor* reported in the seminal 2008 *Nature* paper is a sibling of an *ideal memristor*.

10. There exist *dissipative* circuits made of *generic memristors, inductors*, and *capacitors* which has a ***Hamiltonian***.

3-130

Warm up Reads

1. *L. Chua, V. Sbitnev, H. Kim, "Hodgkin–Huxley Axon is Made of Memristors," International Journal of Bifurcation and Chaos, vol.22, no.03, 2012, p. 1230011 (48 pages).*

2. *L. Chua, V. Sbitnev, H. Kim, "Neurons Are Poised Near the Edge of Chaos," International Journal of Bifurcation and Chaos," vol.22, no.04, 2012, 1250098 (49 pages).*

3. *L. Chua, "Memristor: Remembrance of Things Past", IEEE micro, vol.38, no.5, Sep./Oct. 2018, pp.7-12.*

3-131

Lecture 4

10 More Things You Didn't Know about Memristors

3-132

End Quote for Lecture 3

For every conservation law, there must exist a continuous symmetry.

Emmy Noether

3-133

Appendix

3-134

Universal Formula for Continuous PWL Functions

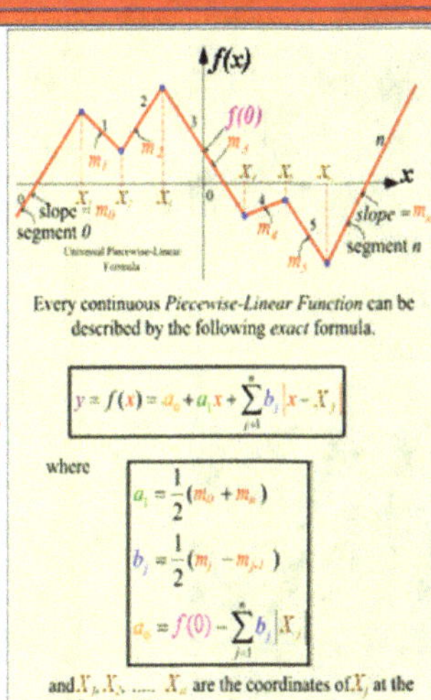

Half Sum Half difference, Then Null Formula

3-135

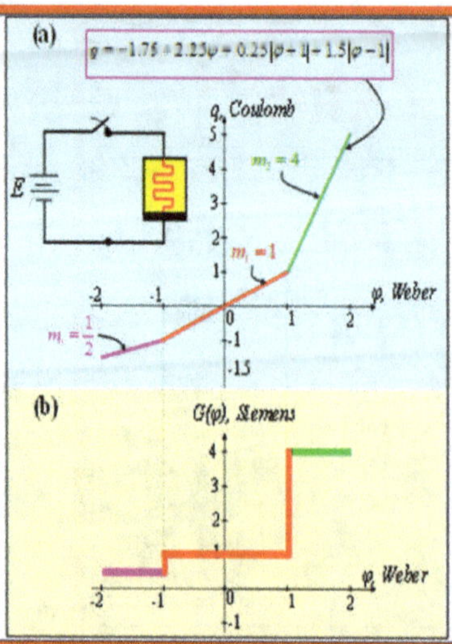

3-136

Our

4 Circuit-Element

Axiomatic

Approach

dates back to

Aristotle !

Aristotle, 350 BC

3-137

3-138

Aristotle's Theory of Matter

3-139

All *matter* consisted of *four elements*:

1. EARTH 2. WATER 3. AIR 4. FIRE

Each of these elements exhibited two of *four fundamental properties* :

1. Moistness 2. Dryness 3. Coldness 4. Hotness

Aristotle's 4 Pairs of Complementary Variables

3-140

- (Dryness, Hotness) ⟹ FIRE
- (Moistness, Hotness) ⟹ AIR
- (Moistness, Coldness) ⟹ WATER
- (Dryness, Coldness) ⟹ EARTH

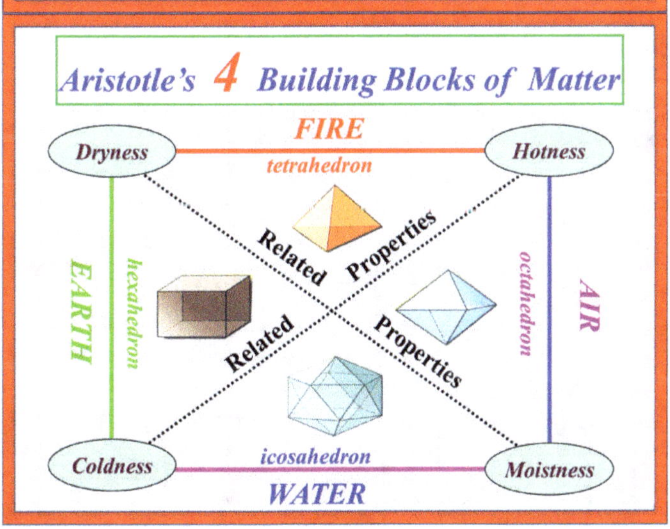

3-141

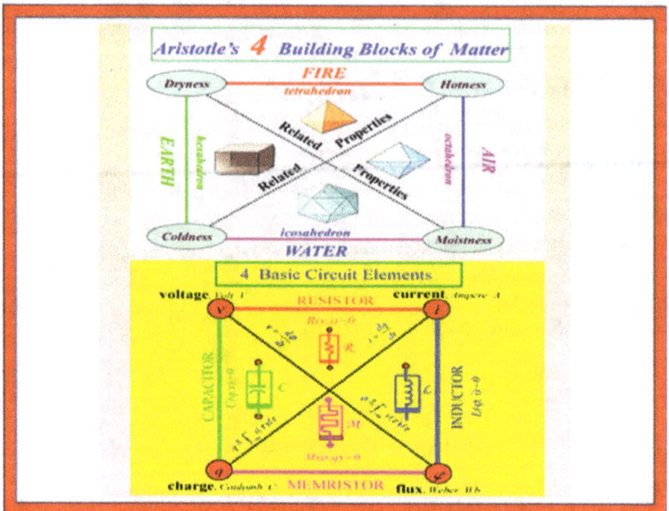

3-142

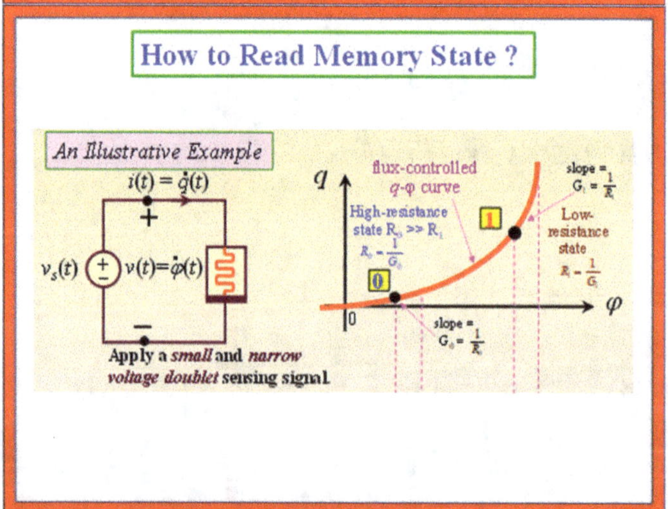

3-143

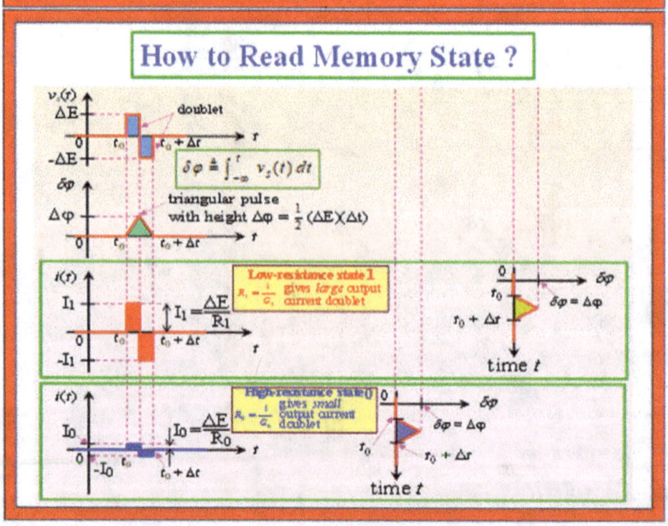

3-144

Companion Media Pack

Supplementary Material

The supplementary material includes the following video(s):

Lecture 1:

Slides 1-132: videos 1, 2, 3, 4

Lecture 3:

Slides 3-125: videos 1, 2

Please follow the instructions below to download the files:

1. Register an account/login at https://www.worldscientific.com.
2. Go to https://www.worldscientific.com/r/11693-vol1-supp to activate the access.
3. Access the materials from:
   https://www.worldscientific.com/worldscibooks/10.1142/11693-vol1#t=suppl

For easy activation, please scan the following QR code to register or log in to activate the access.

For subsequent access, simply log in with the same login details.
For enquiries, please email: sales@wspc.com.sg.

www.ingramcontent.com/pod-product-compliance
Lightning Source LLC
LaVergne TN
LVHW010535300926
849162LV00002B/152